Praise for *The Book of Eels*

"A masterful narrative that is part memoir and part scientific detective story."
—*Smithsonian*

"An unusual and beguiling guide to an unusual and beguiling animal. . . . Svensson's book, like its subject, is a strange beast: a creature of metamorphosis, a shapeshifter that moves among realms. It is a book of natural history, and a memoir about a son and his father. It is also an exploration of literature and religion and custom, and what it means to live in a world full of questions we can't always answer."
—*New Yorker*

"Captivating. . . . Shot through with electric current. The book's deadpan title perhaps undercuts its depth and complexity. Yes, this is a book about eels, those uncanny creatures, but in Svensson's capable hands it is also a book about obsession and mystery, about faith and science, and about the limits of knowledge. . . . Like Annie Dillard and Rachel Carson, Svensson knows the best nature writing is done with emotion and drive."   —*Minneapolis Star Tribune*

"Svensson has, quite stunningly, discovered in the natural and human history of the European eel a metaphor for his father's life and a way to explore questions of knowledge, belief, and faith."
—*Washington Post*

"As much a boon to my mental life as a blow to my social one. For weeks after reading I found myself cornering people at parties to obliterate them with a machine-gun spray of eel facts. But according to *The Book of Eels*, I'm not alone in my eelmania. . . . If you don't think of yourself as someone who might enjoy meditating on eel glory, well, I didn't either, and here I am transcribing my encounter for publication."
—*New York*

"Captivating. . . . *The Book of Eels* is, in the end, not really about eels but about life itself."   —*Wall Street Journal*

"Inspires readers to see eels in a whole new way."
—*Los Angeles Times*, "21 New and Classic Books to Keep You in Touch with the Natural World"

"A beguiling chronicle."   —*Nautilus*

"Enthralling."   —*Colorado Springs Gazette*

"Nature writing at its finest."   —Shelf Awareness

"Without a doubt, the most delicious natural history book of the decade. Svensson's prose effortlessly undulates between his own personal experience and a thousand years of scientific inquiry. But it's his call to conservation—not just of this noble eel but of our memories both personal and cultural—that truly elevates this remarkably poignant work."

—Mark Siddall, former curator and professor,
American Museum of Natural History

"Poses questions about philosophy, the metaphysical, and the spiritual, as well as scientific issues, in a way that will stir readers. This beautifully crafted book challenges us not only to understand eels but also our own selves. Highly recommended."

—*Library Journal* (starred review)

"A wonderful read. The story of the eel is one of the most fascinating on the planet, but equally fascinating is the story Patrik Svensson tells so well here about the mysteries of being."

—Bernd Heinrich, author of *Mind of the Raven*

"Blending a wonderfully evocative and succinct timeline of scientific discoveries about eels with a memoir of his changing relationship with his father, Svensson has produced an extremely readable book on a fish that all have heard of but few (on our side of the pond) have actually seen." —*Booklist*

"An account of the mysterious life of eels that also serves as a meditation on consciousness, faith, time, light and darkness, and life and death. . . . An intriguing natural history." —*Kirkus Reviews*

"Captivating. . . . Nature-loving readers will be enthralled by [his] fascinating zoological odyssey." —*Publishers Weekly* (starred review)

"With lyricism and sharp clarity, Patrik Svensson lets us in on the secret dreamlike world of the eel. As we move deeper into a book that intertwines beautiful nature writing with a moving memoir of a quiet father and a loyal son—as well as healthy doses of philosophical thought from Aristotle to Freud—we get to know one of earth's most unknowable creatures and revel in a life so different than our own." —David Gessner, author of *All the Wild That Remains*

"What an amazing book. About eels!—a haunting and extraordinary creature. When Eugenio Montale wanted a metaphor for the revival of Europe after World War II, he found it in a glimpse of a gold-brown eel in a muddy Ligurian stream, and when Rachel Carson wanted to convey to her readers the intricacy of the relationship between salt- and freshwater, she imagined the migration of eels. Patrik Svensson explores both their mystery and the science that has brought them into focus in the last few decades and made them seem a vivid indicator species for the health of our planet."
—Robert Hass, author of *Summer Snow*

The Book of Eels

THE
BOOK
OF
EELS

*Our Enduring Fascination with
the Most Mysterious Creature in the
Natural World*

PATRIK SVENSSON

Translated from the Swedish by Agnes Broomé

An Imprint of HarperCollins*Publishers*

HarperCollins books may be purchased for educational, business, or sales promotional use. For information, please email the Special Markets Department at SPsales@harpercollins.com.

Ecco® and HarperCollins® are trademarks of HarperCollins Publishers.

Originally published as *Ålevangeliet* in Sweden in 2019 by Albert Bonniers Förlag.

A hardcover edition of this book was published in 2020 by Ecco, an imprint of HarperCollins Publishers.

FIRST ECCO PAPERBACK EDITION PUBLISHED 2021

Designed by Renata De Oliveira
Sea waves illustration by Marzufello/shutterstock

Library of Congress Cataloging-in-Publication Data has been applied for.

ISBN 978-0-06-296882-1 (pbk.)

24 25 26 27 28 LBC 9 8 7 6 5

Later in the same fields
He stood at night when eels
Moved through the grass like hatched fears
—SEAMUS HEANEY

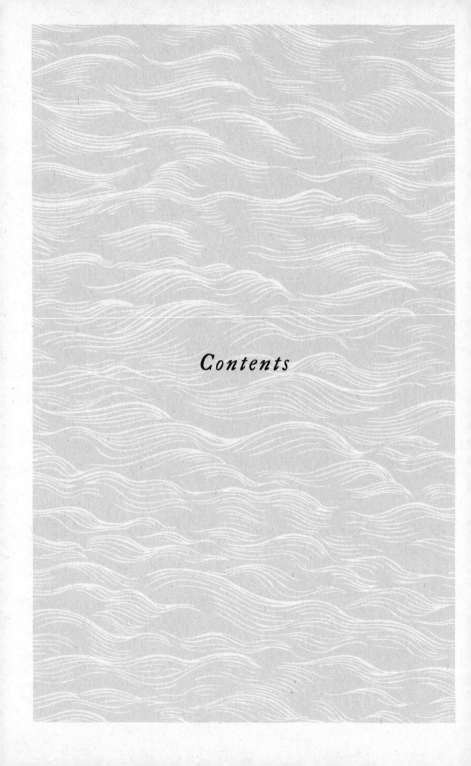

Contents

The Book of Eels

1

The Eel

This is how the birth of the eel comes about: it takes place in
a region of the northwest Atlantic Ocean called the Sargasso
Sea, a place that is in every respect suitable for the creation of
eels. The Sargasso Sea is actually less a clearly defined body of
water than a sea within a sea. Where it starts and where it ends
is difficult to determine, since it eludes the usual measures of
the world. It's located slightly northeast of Cuba and the Baha-
mas, east of the North American coast, but it is also a place in
flux. The Sargasso Sea is like a dream: you can rarely pinpoint
the moment you enter or exit; all you know is that you've been
there.

This impermanence is a result of the Sargasso's being a
sea without land borders; it is bounded instead by four mighty
ocean currents. In the west by the life-giving Gulf Stream; in
the north by its extension, the North Atlantic Drift; in the east
by the Canary Current; and in the south by the North Equa-
torial Current. Two million square miles in size, the Sargasso

Sea swirls like a slow, warm eddy inside this closed circle of currents. What gets in doesn't always have an easy time getting out.

The water is deep blue and clear, in places very nearly 23,000 feet deep, and the surface is carpeted with vast fields of sticky brown algae called *Sargassum*, which give the sea its name. Drifts of seaweed many thousands of feet across blanket the surface, providing nourishment and shelter for myriad creatures: tiny invertebrates, fish and jellyfish, turtles, shrimp, and crabs. Farther down in the deep, other kinds of seaweed and plants thrive. Life teems in the dark, like a nocturnal forest.

This is where the European eel, *Anguilla anguilla*, is born. This is where mature eels breed in the spring and their eggs are laid and fertilized. Here, safe in the darkness of the depths, small larva-like creatures with disturbingly tiny heads and poorly developed eyes spring to life. They're called leptocephalus larvae and have a body like a willow leaf, flat and virtually transparent, only a few millimeters long. This is the first stage of the eel's life cycle.

The gossamer willow leaves immediately set off on their journey. Swept up by the Gulf Stream, they drift thousands of miles across the Atlantic toward the coasts of Europe. It's a journey that can take as long as three years; during this time, each larva slowly grows, millimeter by millimeter, like a gradually inflating balloon, and when at last it reaches Europe, it undergoes its first metamorphosis, transforming into a glass eel. This is the second stage of the eel's life cycle.

Glass eels are, much like their willow leaf former selves, almost entirely transparent, two to three inches in length, elongated and slithery, transparent, as though neither color nor sin has yet to take root in their bodies. They look, in the words of the marine biologist Rachel Carson, like "thin glass rods, shorter than a finger." Frail and seemingly defenseless, they are considered a delicacy by, among other people, the Basques.

When a glass eel reaches the coasts of Europe, it will usually travel up a brook or river, adapting almost instantly to a freshwater existence. This is where it undergoes yet another metamorphosis, turning into a yellow eel. Its body grows serpentine and muscular. Its eyes remain relatively small, with a distinctive dark center. Its jaw becomes wide and powerful. Its gills are small and almost completely concealed. Thin, soft fins stretch along the entirety of its back and belly. Its skin finally develops pigment, coloring it shades of brown, yellow, and gray, and it becomes covered in scales so tiny they can be neither seen nor felt, like an imaginary armor. If the glass eel is tender and fragile, the yellow eel is strong and sturdy. This is the third stage of the eel's life cycle.

The yellow eel is able to move through the shallowest, most overgrown waters as well as the swiftest currents. It can swim through murky lakes and up tranquil streams, up wild rivers and through lukewarm ponds. When needed, it can pass through swamps and ditches. It doesn't let circumstance stand in its way, and when all aquatic possibilities have been exhausted, it can take to dry land, slithering through moist brush and grass in pushes toward new waters that can last for hours. The eel is, thus, a fish that transcends the piscine condition. Perhaps it doesn't even realize it is a fish.

It can migrate thousands of miles, unflagging and undaunted, before it suddenly decides it's found a home. It doesn't require much of this home; the environs are something to adapt to, to endure and get to know—a muddy stream or lake bed, preferably with some rocks and hollows to hide in, and enough food. Once it has found its home, it stays there, year after year, and normally wanders within a radius of only a few hundred yards. If relocated by external forces, it will invariably return as quickly as it can to its chosen abode. Eels caught by researchers, tagged with radio transmitters, and released many miles from their point of capture have been known to return to where they were first found within a week or two. No one knows exactly how they find their way.

The yellow eel is a solitary creature. It usually lives out the active phase of its life alone, letting the passing seasons dictate its activities. When the temperature drops, it can lie motionless in the mud for long periods, utterly passive, and at times entangled with other eels like a messy ball of yarn.

It is a nocturnal hunter. At dusk, it emerges from the sediment and starts looking for food, eating whatever it can find. Worms, larvae, frogs, snails, insects, crayfish, fish, as well as mice and baby birds when given the chance. It is not above scavenging.

In this way, the eel lives out the greater part of its life in a brownish-yellow guise, alternating between activity and hibernation. Seemingly lacking any sense of purpose, other than in its daily search for food and shelter. As though life was first and foremost about waiting and its meaning found in the gaps or in an abstract future that can't be brought about by any means other than patience.

And it's a long life. An eel that successfully avoids illness and calamity can live for up to fifty years in one place. There are Swedish eels who have made it past eighty in captivity. Myths and legends tell of eels living to a hundred or more. When an eel is denied a way to achieve its main purpose in life—procreation—it seems able to live forever. As though it could wait until the end of time.

But at some point in its life, usually after fifteen to thirty years, a wild eel will suddenly decide to reproduce. What triggers this decision, we may never know, but once it has been made, the eel's tranquil existence ends abruptly and its life takes on a different character. It starts making its way back to the sea while simultaneously undergoing its final metamorphosis. The drab and indeterminate yellowish-brown of its skin disappears, its coloring grows clearer and more distinct, its back turns black and its sides silver, marked with stripes, as though its entire body changes to reflect its newfound determination. The yellow eel becomes a silver eel. This is the fourth stage of the eel's life cycle.

When autumn rolls out its protective darkness, the silver eels wander back out into the Atlantic and set off toward the Sargasso Sea. And as though through deliberate choice, the eel's body adapts to the conditions of the journey. Only now do its reproductive organs develop; its fins grow longer and more powerful to help propel it; its eyes grow larger and turn blue to help it see better in the depths of the ocean; its digestive system shuts down; its stomach dissolves—from now on, all the energy it needs will be taken from existing fat reserves—its body fills with roe or milt. No external interference can distract the eel from its goal.

It swims as much as thirty miles a day, sometimes as deep as three thousand feet below the surface; we still know very little about this journey. It may make the trip in six months or it may stop for winter. It has been shown that a silver eel in captivity can live for up to four years without any nourishment at all.

It's a long, ascetic journey, undertaken with an existential resolve that cannot be explained. But once an eel reaches the Sargasso Sea, it has, once again, found its way home. Under swirling blankets of seaweed, its eggs are fertilized. And with that, the eel is done, its story complete, and it dies.

2

By the Stream

My father taught me to fish for eel in the stream bordering the fields of his childhood home. We drove down at dusk in August, taking a left off the main road to cross the stream and turning onto a small road that was little more than a tractor path in the dirt winding down a steep slope and then moving parallel with the water. On our left were the fields, the golden wheat brushing against the side of our car; on our right, the quietly hissing grass. Beyond it, the water, around twenty feet wide, a tranquil stream meandering through the greenery like a silver chain glinting in the last slanted rays of the setting sun.

We drove slowly along the rapids, where the stream rushed in a startled fashion between the rocks and past the twisted old willow tree. I was seven years old and had already gone down this same road many times before. When the tracks ended in a wall of impenetrable vegetation, Dad turned off the engine and everything went dark and still, aside from the murmur of the stream. We were both wearing wellies and greasy vinyl waders, mine yellow and his orange, and we took two black buckets full of fishing gear, a flashlight, and a jar of worms from the trunk and set off.

Along the bank of the stream, the grass was wet and impenetrable and taller than me. Dad took the lead, forging a path; the vegetation closed like an arch above me as I followed. Bats flitted back and forth above the stream, silent, like black punctuation marks against the sky.

After forty yards, Dad stopped and looked around. "This'll do," he said.

The bank was steep and muddy. If you missed your step, you ran the risk of falling over and sliding straight into the water. Twilight was already falling.

Dad held the grass back with one hand and carefully walked down on a diagonal, then turned around and held his other hand out to me. I took it and followed with the same practiced caution. Down by the water's edge, we trampled out a small ledge and set down our buckets.

I imitated Dad, who was mutely inspecting the water, following his eyes, imagining I saw what he saw. There was, of course, no way of knowing whether this was a good spot. The water was dark, and here and there stands of reeds stuck out of it, waving menacingly, but everything below the surface was hidden from us. We had no way of knowing, but we chose to have faith as from time to time a person must. Fishing is often about exactly that.

"Yes, this'll do," Dad repeated, turning to me; I pulled a spiller from the bucket and handed it to him. He pushed the stake into the ground and quickly gathered up the line, picked up the hook, and gingerly pulled a fat worm out of the jar. He bit his lip and studied the worm in the flashlight; after putting it on the hook, he held it up to his face and pretended to spit on it for luck, always twice, before throwing it into the water with a sweeping motion. He bent down and touched the line, making sure it was taut and hadn't traveled too far in the current. Then he straightened back up and said "All right," and we climbed back up the bank.

What we called spillers were really something else, I suppose. The word *spiller* usually denotes a long fishing line with many hooks and sinkers. Our version was more primitive. Dad made them by sharpening one end of a piece of wood with an ax. Then he cut a length of thick nylon line, about fifteen feet, and tied one end to the wooden stake. He made the sinkers by pouring melted lead into a steel pipe and letting it set before cutting the pipe into short pieces that he would then drill a hole through. The sinker was placed about a hand's length from the end of the line and the fairly sizable single hook right fastened at the end. The stake was hammered into the ground, the hook with the worm rested on the streambed.

We would bring ten or twelve spillers, which we'd bait and throw in, one after the other, approximately thirty feet apart. Up and down the steep bank, the same laborious procedure each time and the same well-rehearsed hand-holding, the same gestures and the same spitting for luck.

When the last spiller had been set up, we went back the same way, up and down the bank, checking each one again. Carefully testing each line to make sure there hadn't been a bite already and then standing around for a minute in silence, letting our instinct convince us that this was good, that something would happen here if we just gave it some time. By the time we'd checked the last one, it would be completely dark—the silent bats visible now only when they swooped through the shaft of moonlight—and we climbed up the bank one final time, walked back to the car, and drove home.

I CAN'T RECALL US EVER TALKING ABOUT ANYTHING OTHER THAN eels and how to best catch them, down there by the stream. I can't remember us speaking at all.

Maybe because we never did. Because we were in a place where the need for talking was limited, a place whose nature was best enjoyed in silence. The reflected moonlight, the hissing grass, the shadows of the trees, the monotonous rushing of the stream, and the bats like hovering asterisks above it all. You had to be quiet to make yourself part of the whole.

It could, of course, also be because I remember everything wrong. Because memory is an unreliable thing that picks and chooses what to keep. When we look for a scene from the past, it is by no means certain that we end up recalling the most important or the most relevant; rather, we remember what fits into the preconceived image that we have. Our memory paints a tableau in which the various details inevitably complement one another. Memory doesn't allow colors that clash with the background. So let's just say we were silent. In any case, I don't know what we might have talked about if we did.

We lived just a mile or two from the stream; when we got home late at night, we would pull off our wellies and waders on the front steps, and I would go straight to bed. I'd fall asleep quickly, and just after five in the morning, Dad would wake me up again. He didn't need to say much. I got out of bed straight away, and we were in the car a few minutes later.

Down by the stream, the sun was rising. Dawn colored the lower edge of the sky a deep orange, and the water seemed to rush by with a different sound, clearer, brighter, as though it had just woken up from a deep sleep. Other sounds could be heard all around us. A blackbird warbling, a mallard entering the water with a clumsy splash. A heron flying silently over the stream, peering down with its large beak like a raised dagger.

We walked through the damp grass and stomped our way sideways down the bank to the first spiller. Dad waited for me, and together we studied the taut line, looking for signs of activity under the surface. Dad bent down and put his hand to the nylon. Then he straightened back up and shook his head. He pulled the line in and held up the hook for me to see. The worm was gone, probably stolen by crafty roaches.

We moved on to the next spiller, which was also empty. As was the third. Approaching the fourth one, however, we could see the line had been dragged into a stand of reeds; when Dad pulled on it, it was stuck. He muttered something inaudible. Grabbed the line with both hands and tugged a bit harder, to no avail. The current might have carried the hook and sinker into the reeds. But it might also have been that an eel had swallowed the hook and gotten itself and the line caught up in the plant stalks and was now lying there, biding its time. If you held the line taut in your hand, you could sometimes feel tiny movements, as though whatever was stuck below the surface on the other end was bracing itself.

Dad coaxed and pulled, bit his lip and cursed helplessly. He knew there were only two ways out of this situation and that both had its losers. Either he managed to dislodge the eel and pull it up, or he could cut the line and leave the eel where it was, tangled in the reeds with the hook and heavy sinker like a ball and chain.

This time, there seemed to be no other option. Dad took a few steps to the side, trying a different angle, pulling so hard the nylon stretched like a violin string. Nothing worked.

"Nope, no luck," he said at length and tugged as hard as he could, breaking the line in two with a loud snap.

"Let's hope it makes it," he said, and we moved on, climbing up and down the bank.

At the fifth spiller, Dad bent down and tentatively touched the line. Then he straightened up and stepped aside. "You want to take this one?" he said.

I grabbed the line and pulled on it gently and could immediately feel the strength that answered back. The same force that Dad had felt with just his fingertips. I had time to realize that the feeling was familiar, then I pulled a bit harder and the fish began to move. "It's an eel," I said out loud.

An eel never tries to rush, as a pike might; it prefers slithering sideways, which creates a kind of undulating resistance. It's surprisingly strong for its size and a good swimmer, despite its tiny fins.

I reeled it in as slowly as I could, without letting the line slacken, as though savoring the moment. But it was a short line, and there were no reeds for this eel to hide in; before long, I pulled it out of the water and saw its shiny yellowish-brown body twisting in the early-morning light. I tried to grab it behind its head, but it was virtually impossible to hold. It wrapped itself around my arm like a snake, up past my elbow;

I could feel its strength like a static force more than move-ment. If I dropped it now, it would escape through the grass and back into the water before I could get a secure hold.

In the end, we got the hook out and Dad filled the bucket with water from the stream. I slipped the eel in, and it imme-diately started swimming around and around the inside; Dad put his hand on my shoulder, said it was a beauty. We moved on to the next spiller, stepping lightly up the bank. And I got to carry the bucket.

3

Aristotle and the Eel Born of Mud

There are circumstances that force us to choose what to believe. The eel is one such circumstance. If we choose to believe Aristotle, all eels are born out of mud. They simply appear, as though out of thin air, in the sediments at the bottom of the sea. In other words, they're not created by other eels reproducing, by the union of reproductive organs and the fertilization of an egg.

Most fish, Aristotle wrote in the fourth century BCE, do, of course, lay eggs and breed. But the eel, he explained, is an exception. It is neither female nor male. It neither lays eggs nor mates. Eels do not give life to other eels. The spark of their life comes from somewhere else.

Aristotle suggested: Study a small pond with a tributary during a period of drought. When the water has evaporated and all the mud and muck has dried out, there is no life at all to be found on its hardened bottom. No life can be sustained there, much less a fish. But when the first rain comes and the water slowly returns, something incredible happens. Suddenly, the pond is once more full of eels. Suddenly, they're just there. The rainwater brings them into existence.

Aristotle's conclusion was that eels simply spring into being, like a slithering, enigmatic miracle.

Aristotle's interest in eels is not entirely unexpected. He was interested in all forms of life. He was, of course, a thinker and theoretician and the man who, along with Plato, laid the foundation for all Western philosophy; but more than that, he was a scientist, at least by the standards of his age. It's often said that Aristotle was the last person to "know it all"; or in other words, he was the last person to possess all the knowledge accumulated by humanity. And, among other things, he was ahead of his time when it came to observing and describing nature. His great work *Historia Animalium* (*The History of Animals*) was a first attempt, more than two thousand years before Linnaeus, to systematically categorize the animal kingdom. Aristotle observed and described a wide range of animals and what differentiated one from another. What they looked like, their body parts, coloring, and shape, how they lived and procreated, what they ate, their behaviors. Modern zoology grew out of the *Historia Animalium*; it remained a standard work in the natural sciences well into at least the seventeenth century.

Aristotle grew up in Stagira on Chalcidice: a peninsula ending in three narrow spits of land that jut out into the Aegean Sea, like a hand with three fingers. His life was one of privilege, with a father who was physician to the Macedonian king; he received a good education, and his father likely envisioned a future as a doctor for his son. But Aristotle was orphaned at a young age. His father died when he was about ten, his mother probably before that. He was taken in by a relative and at seventeen was sent to Athens to study at the finest school in antiquity, the Platonic Academy. A young man, alone in a strange city, curious and brilliant and with a passion for understanding the world that can be comprehended only by those whose own roots have been severed. He studied at Plato's feet in Athens for twenty years and in many respects came to be his equal. When Plato died and Aristotle was not appointed the new head of the Academy, he relocated to the island of Lesbos. It was there that he began to study animals and nature in earnest. Perhaps that was also where he first started thinking about how eels came to be.

Not much is known about Aristotle's scientific method. He didn't keep notes on his observations and dissections. He gave confident and detailed accounts of his discoveries and insights, but rarely said anything about how he had come to them. Nevertheless, we can be almost entirely certain that he personally performed many of the dissections that form the basis of *Historia Animalium*. Crucially, it seems clear he spent much of his time studying aquatic life-forms, and primarily the eel. If nothing else, his writings on what is hidden inside the eel, about the relative placement of its organs and the construction of its gills, are particularly copious and detailed.

Where the eel is concerned, he also often disagreed with other scientists whose names have been lost to posterity, as though the eel was already, at that time, a source of speculation, contradictory opinions, and conflict. Aristotle insisted categorically that eels never carry eggs in their bodies, declaring that anyone who claimed otherwise simply had not studied eels closely enough. There can be no doubt this is so, he wrote, because when you cut open an eel, not only will you not find eggs, you will also not find any organs for producing or transporting eggs or milt. Nothing about the eel's existence explains how it is brought to life. He also stated that anyone claiming the eel gives birth to live young had been misled by his ignorance and that his opinions were not based on fact. Aristotle also made short shrift of those scientists who claimed eels could be sexed, pointing to the male head as being larger than that of the female. They had simply mistaken interspecies variation for sexual variation.

Aristotle had studied eels, that much is clear. Maybe on Lesbos, maybe in Athens. He had dissected them and studied their internal organs, had looked for eggs and reproductive organs and an explanation as to how they procreate. He had probably handled a great many eels, scrutinizing them, pondering what kind of creatures they were. And he had reached the conclusion that the eel is a thing unto itself.

The approach to understanding animals and nature developed by Aristotle would eventually come to shape—virtually single-handedly—both modern biology and the natural sciences, and thus all subsequent attempts to understand the eel. It was above all empirical. Nature can be described through systematic observation, Aristotle claimed, and only through correct description can it be understood.

It was a radical approach and, in every respect, a successful one. Many of Aristotle's observations were surprisingly precise, not least considering they were made long before the field of zoology even existed as a concept. His knowledge was way ahead of his time, particularly when it came to aquatic species. He explained and described, for example, the anatomy and reproduction of octopuses in a way that modern zoology was able to verify only in the nineteenth century. And with regards to the eel, Aristotle claimed, correctly, that it can move between freshwater and saltwater, that it has unusually small gills, and that it is nocturnal, hiding in deeper water during the day.

But the eel was also a subject about which Aristotle made an unusual number of obviously outlandish claims. Despite his systematic method based on observation, he never did manage to understand the eel. He wrote that eels eat grass and roots and sometimes even mud. He wrote that it has no scales. He wrote that it lives for seven or eight years and that it can survive for five or six days on land and even longer if the wind blows from the north. And, as already mentioned, he asserted that eels do not have biological sex and that they are created from nothing. The first embodiment of the eel, Aristotle concluded, is in fact a small maggot-like creature, a kind of earthworm that is spontaneously and without the involvement of any other living thing generated from mud. This worm can spring to life in both seas and rivers, especially where there is plenty of decomposing vegetation, and it prefers shallow marshes or beds of seaweed where the sun warms the water. "There can be no doubt about this being so," Aristotle writes, and then wraps up his discussion. "Enough about the reproduction of the eel."

ALL KNOWLEDGE COMES FROM EXPERIENCE. THAT WAS ARISTOTLE'S
first and most fundamental insight. Any study of life must be
empirical and systematic. Reality must be described as it is
perceived by our senses. First, one establishes *that* something
is; then one can focus on the question of *what* it is. And only
when one has collected all the facts about *what* something is,
is it possible to approach the metaphysical question of *why* it
is the way it is. That is also the insight that has served as the
basis for most attempts to gain a scientific understanding of
the world since Aristotle's time.

But why is it that the eel was able to slither out of Aristotle's
grasp? That is the question that seems impossible to answer.
No matter how meticulously and systematically he studied the
eel, he reached conclusions that now appear almost absurdly
unscientific. And that's what makes the eel unique. Science
has come up against many mysteries, but few have proven as
intractable and difficult to solve as the eel. Eels have turned
out to be not only uncommonly difficult to observe—due to
their strange life cycle, their shyness, their metamorphoses,
and their roundabout approach to reproduction—but also
secretive in a way that comes across as deliberate and preordained. Even when successful observation is possible, even
when you get really close, the eel seems to pull away. Given
the inordinate amount of time so many people have spent
studying and trying to understand the eel, we should, simply
put, know more than we do. That we don't is something of a
mystery. Zoologists call it "the eel question."

Aristotle may have been one of the first to document his
misapprehensions about the eel, but he was, as we know, not
the last. The eel has continued to elude scientific study into

our modern era. Any number of prominent researchers, as well as amateurs with varying degrees of enthusiasm, have studied the eel without ever really understanding it. Some of the most noted names in the history of natural science have tried in vain to find the answer to the eel question. It's as though their senses were not enough in themselves. Somewhere in the darkness and mud, the eel has managed to hide away from human knowledge. When it comes to eels, an otherwise knowledgeable humanity has always been forced to rely on faith to some extent.

In the olden days, a distinction was likely often made between eels and other fish. The eel was a creature apart, with its appearance and behavior, its invisible scales and barely visible gills and ability to survive out of water. It was different enough to make many people believe it was in fact an aquatic snake or amphibian. Homer himself seemed to distinguish eels from fish. After Achilles kills Asteropaios in the *Iliad* he "let him lie where he was on the sand, with the dark water flowing over him and the eels and fishes busy nibbling and gnawing the fat that was about his kidneys." Today, the question is still asked from time to time: Is the eel really a fish?

This uncertainty about the fundamental nature of the eel has often led to some distance between us and them. People have found eels frightening or disgusting. They're slimy and slithery, look like snakes and are said to eat human bodies; they move surreptitiously, in the dark and the mud. The eel is alien, unlike other animals, and regardless of how ubiquitous it has been, in our lakes and rivers and on our tables, it has always remained a stranger in some respects.

The most abiding and debated mystery about the eel has

been its method of reproduction. It's only in the past century that we've been able to give a reasonable, if not conclusive explanation. For a long time, many people simply chose to believe Aristotle and his theory about worms springing into being spontaneously from mud. Others sided with the natural philosopher Pliny the Elder, who perished in the eruption of Mount Vesuvius in AD 79, and who claimed that the eel reproduced by rubbing itself against rocks, which freed particles from its body that in turn became new eels. Some believed the Greek author Athenaeuss, who in the third century explained that the eel secreted a kind of fluid that sank into the mud and became new life.

More or less fanciful theories have been proposed throughout history. The ancient Egyptians were convinced eels sprang to life from nothing when the sun warmed the waters of the Nile. In various parts of Europe, it was thought that eels were born from decomposing vegetation on the seafloor or grew out of the rotting cadavers of other, dead eels. Some believed eels were born of sea-foam or created when the rays of the sun fell on a certain kind of dew that covered lakeshores and riverbanks in the spring. In the English countryside, where eel fishing was popular, most people adhered to the theory that eels were born when hairs from horses' tails fell into water.

Many of the different theories about the birth of the eel clearly revolve around a common notion. That is to say, the notion that life can spring from something seemingly lifeless, a minute echo of the birth of the universe itself. A mosquito born of a speck of dust, a fly born out of a piece of meat, an eel born of mud—such an idea has been commonly referred to as *spontaneous generation* and has historically been a widespread idea, particularly before the invention of the microscope. Simply put, people believed what they could see, so if you were looking at a piece of rotting meat and suddenly saw maggots crawl out of it, without having observed any flies or fly eggs, how could you conclude anything but that the larvae had been created out of thin air? In the same way, no human has observed procreating eels, and as far as anyone could tell, they had no reproductive organs.

The idea of spontaneous generation leads back, of course, to the creation of everything, to the creation of life itself. If there was in fact once a beginning, when life sprung into existence from nothing (whether you attribute it to divine intervention or some other factor), it may not have been so outlandish to assume that such spontaneous generation could be repeated.

How it supposedly happened has been explained several ways. In Genesis, there is mention of a "wind from God" sweeping across the barren, desolate earth, creating not only light and land and plants but all the animals, too. The ancient philosophers known as the Stoics spoke of *pneuma*, the breath of life, a combination of air and heat needed for the existence of both living bodies and the soul. The underlying premise is a belief that nonliving matter can be turned into living matter, that the living and the dead are in fact dependent on one another and that some kind of life can exist in something seemingly dead. When the eel could not be understood or explained, that kind of thinking was clearly close at hand; the eel became a reflection of the deeper mystery of life's origins.

What makes eels special, however, is that we're still forced to rely on faith to some extent as we try to understand them. We may think we now know everything about the life and reproduction of the eel—its long journey from the Sargasso Sea, its metamorphoses, its patience, its journey back to breed and die—but even though that is all probably true and correct, much of it is nevertheless still based on assumption.

No human has ever seen eels reproduce; no one has seen an eel fertilize the eggs of another eel; no one has managed to breed European eels in captivity. We think we know that all eels are hatched in the Sargasso Sea, since that's where the smallest examples of the willow leaf–like larvae have been found, but no one knows for certain why the eel insists on reproducing there and only there. No one knows for certain how it withstands the rigors of its long return journey, or how it navigates. It's thought all eels die shortly after breeding, since no living eels have ever been found after breeding season, but then again, no mature eel, living or dead, has ever been observed at their supposed breeding ground. Put another way, no human has ever seen an eel in the Sargasso Sea. Nor can anyone fully comprehend the purpose of the eel's many metamorphoses. No one knows how long eels can live for.

In other words, more than two thousand years after Aristotle, the eel remains something of a scientific enigma, and in many ways, it has become a symbol of what is sometimes referred to as the metaphysical. As it happens, metaphysics can also be traced back to Aristotle (though the concept was named only after his death). It is a branch of philosophy that is concerned with what exists outside, or beyond, objective nature, beyond what we can observe and describe with the help of our senses.

Metaphysics is not necessarily concerned with God. It is, rather, an attempt to describe the true nature of things, the *whole* of reality. It claims there's a difference between existence per se and the characteristics of that existence. It also claims the two questions are separate. The eel *is*. Existence comes first. But *what* it is, is a completely different matter.

I like to think that's why the eel has continued to be a source of fascination. Because that intersection between knowledge and faith, where knowledge is incomplete and therefore allowed to contain both fact and traces of myth and imagination, is compelling. Because even people who trust in science and an orderly natural world sometimes want to leave a small, small opening for the unknowable.

If you are of the opinion that the eel should be allowed to remain an eel, it follows that you have to allow it to remain a mystery, to some degree. For now, at least.

AND THE EEL DID REMAIN A MYSTERY. IS IT A FISH OR SOMETHING else entirely? How does it reproduce? Does it lay eggs or give birth to live young? Is it asexual? Is it hermaphroditic? Where is it born and where does it die? For centuries after Aristotle, the eel was the subject of countless theories, and every attempt to understand it was inevitably suffused with mystique. During the Middle Ages, two theories in particular were popular, often in combination: one that said the eel was viviparous, which is to say it gives birth to live young; another that said the eel was hermaphroditic, both male and female.

With the resurgence of natural science in the seventeenth century, the eel question became the subject of more methodical inquiry. Aristotle's methods were revived—especially his insistence on the need to systematically observe nature—and as a consequence, our view of the world, and the eel, changed.

Yet even so, it would be a long time before the questions about the eel began to find answers. Aristotle had strongly argued against the theory that the eel was viviparous, but it now grew more popular. It was advocated by, among others, the English author Izaak Walton, who in 1653 published the world's first commercially successful book on fishing, *The Compleat Angler.* The eel, he claimed, is viviparous and gives birth to live young, but it is also sexless. New eels were generated inside older ones without conception.

Then the Italian physician and scientist Francesco Redi, of Pisa, published the first evidence-based critique of the concept of spontaneous generation. In 1668, his experiments on flies demonstrated that eggs and fertilization are required to create life. *Omne vivum ex ovo*, he concluded. All life stems from the egg. He also studied eels and managed to show that the tiny wormlike creatures sometimes found inside eels, which some had taken to be unborn young, were in fact more likely parasites. The eel was in all probability not viviparous, Redi wrote, though he never did manage to find any reproductive organs or eggs and was therefore unable to give a definitive answer to the question of *how* the animal really reproduces.

It was in this context that a sensation landed on a table at the University of Padua in Italy. The year was 1707, and a surgeon by the name of Sancassini had visited an eel fishery in Comacchio on Italy's east coast. There he had spotted an eel so big and fat, he had felt compelled to pick up his scalpel and cut it open. Inside the eel, he had found something that looked very much like reproductive organs, and something that resembled eggs.

He sent the dissected eel to his friend Antonio Vallisneri, a professor of natural history in Padua. Vallisneri, a sworn enemy of the notion that life can spring from nothing, was justifiably excited and sent the eel on to the University of Bologna, where many of the most prominent scientists of his day were to be found.

The Comacchio eel breathed new life into the question of the eel's reproduction, the solving of which for a while became the central object of scientific efforts during the Enlightenment. The eel itself was not, however, as well received as Vallisneri had hoped. What had really been found, after all? Granted, it might look like reproductive organs and eggs, but how could anyone know for sure? In order to consider something proved, systematic observation and further study were required; instead of enlightenment, the eel prompted a moderate flare-up in academic debate. A renowned anatomy professor, Antonio Maria Valsalva, was of the opinion that what Vallisneri wanted to call reproductive organs and eggs were in all likeliness common, unsensational fatty tissue. Someone else claimed it was probably a collapsed swim bladder. The doubts provoked squabbling within the scientific community. A professor by the name of Pietro Molinelli offered a reward to anyone who could produce an eel with verifiable eggs inside. He did receive one promising specimen, until it was discovered that the fisherman who had provided the eel in the hopes of pocketing the reward had crammed it full of roe from a completely different species of fish.

And so the Comacchio eel became something of an academic legend—but the eel question remained unanswered. What had in fact been found was never fully agreed on. And in Sweden, Carl Linnaeus, who in 1758 gave the European eel its scientific name, came to the perhaps more convenient conclusion that the eel probably gives birth to live young.

It would take seventy more years after Vallisneri's insight before there was another breakthrough in the eel question. In an almost uncanny instance of repetition, another eel, also caught near Comacchio, ended up on a table at the University of Bologna. This time, the table belonged to Carlo Mondini, a professor of anatomy who would later become famous for his description and naming of a deformity in the human ear that causes deafness. Mondini examined the eel and wrote a now classic treatise, in which the reproductive organs and eggs of a sexually mature female eel were for the first time described with a measure of scientific accuracy. The original Comacchio eel, the one Antonio Vallisneri had sent to Bologna seventy years prior, had, according to Mondini, been misunderstood. By comparing his own findings with those of his predecessors, he was able to establish that what had been found in that eel could with some degree of certainty be said to be a collapsed swim bladder. But this new eel was the real thing. The folds inside it really were its reproductive organs, and the tiny droplet-shaped objects inside really were eggs.

It was 1777, and the question of what the eel is could finally be said to have been provisionally answered. If eels could possess reproductive organs, and be shown to produce eggs, at least that demonstrated they weren't the products of spontaneous generation. The eel still remained a mystery in many respects, but at least a mystery with a degree of anchoring in the observable, describable world. Mondini's discovery brought eels and humans a little bit closer. Now all that was missing was the second half of the equation.

4

*Looking into
the Eyes of an Eel*

My father liked eel fishing for several reasons. I don't know
which was the most important.

What I do know is that he liked it down by the stream.
He liked the magical, overgrown environment, the quietly
rushing water, the willow tree, and the bats. It was only a
few hundred yards from his childhood home, a farm with a
main house and stables from which a narrow gravel path led
down the gentle slope toward the stream. My father had run
up and down that path as a child, to go fishing or swimming.
The stream had constituted the metaphorical outer limit of
his world. He had crept through the tall grass by the water's
edge, catching live mice, which he'd put in his pocket and
bring home to use for slingshot target practice in the yard. He
had skated on the frozen overflows in the winter. In the summer, he had been able to hear the sound of the rapids when he
was kneeling in the fields, thinning beets or picking potatoes.

The stream represented his roots, everything familiar he always returned to. But the eels moving through its depths, occasionally revealing themselves to us, represented something else entirely. They were, if anything, a reminder of how little a person can really know, about eels or other people, about where you come from and where you're going.

I also know Dad liked eating eel. In the summer, when there had been a lot of fishing, he would happily have eel several times a week. He would usually eat it with potatoes and melted butter. Mum did the cooking, taking the skinned, cleaned eel we provided and cutting it up into four-inch pieces that she breaded and fried in butter with a pinch of salt and pepper. I liked to watch. Every time she placed the fish in the hot pan, something incredible happened. The bits of eel moved. They twitched spasmodically in the searing heat. As though there were still life left in them.

I would stand next to my mother and watch in wonder. A body that had just been alive but was now dead, cut into pieces even. And yet, it moved! If death meant motionlessness, could it really be said that the eel was dead? If death robs us of the ability to feel, how come the eel could still feel the heat in the pan? There was no heart beating, but there was some kind of life in it. I wondered where to draw the line between life and death.

Later on, I read that octopuses have myriad nerve endings in their limbs. There are in fact more nerve cells in an octopus's limbs than in its brain, and each prehensile arm is also a nerve center, independent of the central brain in the animal's head. It's as though octopuses have small but autonomous brains at the end of each arm—which is to say that each one can act of its own volition. An octopus can, for example, both taste and feel with its arms, and some species even have photosensitive cells in their limbs, which give them some degree of vision. But what's more; if you cut off an octopus's arm, it doesn't just continue to move, it acts almost like an independent creature. Throw it a piece of food and it will seize it and try to feed the head to which it's no longer attached.

I'd seen similar behavior in eels. I had cut one's head off and watched the rest of the body slither away as though trying to save itself. It continued to move for minutes without a head. To the eel, death seemed relative.

For my part, I ate eel only if I had to, not because I felt sorry for them but because I didn't like the taste. The greasy, slightly gamey flavor made me nauseated. But Dad loved eel. He ate it with his hands, gnawing the bones clean and licking the grease off his fingers. "So fatty and tasty," he'd say. If he didn't eat the eel fried, he ate it boiled. The same four-inch pieces were placed in a pot of salted water with allspice and bay leaves. The meat turned completely white with an oily slickness to it. I liked boiled eel even less than fried.

I didn't, however, mind taking care of the fish we'd caught. When we returned from the stream in the early morning, we brought the eels in that black bucket full of stream water. We filled an even bigger bucket with clean water and transferred the eels. Then we let them sit there for a few hours, sometimes all day. We might change the water at some point.

I would often go outside to have a look at them. My mother ran a day care center, so our house was full of children; I used to take them out to the garage, where the bucket was. I'd poke the eels, trying to make them swim around. I'd demonstrate how to hold them, with your index and middle finger on both sides of the body and your thumb like a hook underneath. I'd pick the eels up and let them writhe and flex in the air. They could lie completely motionless in the bucket, as though dead or paralyzed, but as soon as I picked one up, it would become suddenly violently powerful, wrapping itself around my arm. I'd reek of eel slime. I never let the other children touch the eels.

As evening came on, we'd kill the eels, a brutal spectacle. Dad would pick up an eel and hold it down against a table, grab his fishing knife, and ram the sharp point straight through its head. The eel would writhe in rapid convulsions, tensing its body as though it were one big muscle. When it calmed down a little, Dad would pull the knife out and put the eel on a three-foot-long wooden board. He'd secure it to the board with a five-inch nail hammered through its head so the eel hung suspended as if on a crucifix. With his knife, he would then make an incision, all the way around the body, right below the head.

"Let's take off its pajamas," Dad would say and hand me a pair of pliers. I'd get a firm grip on the edge and pull the skin off in one long, fluid motion. It was blueish on the inside. Like a child's pajamas. Sometimes the body would still be undulating slowly, sluggishly.

We opened the eel and cleaned out the innards, cut the head off, and then it was done. If it was a big eel, we sometimes weighed it, but they were almost always roughly the same size, between one and two pounds. The girth and color would vary slightly; some were paler and others a darker yellowish-brown, but on the whole, they looked remarkably alike. In all the years we fished for eels, we never caught one that weighed more than a little over two pounds. Granted, we considered that gigantic, but we also knew there were supposedly eels that weighed as much as four or five. These were the eels my father dreamed of. He'd read in the paper about an amateur fisherman transforming himself into an expert at catching big eels.

"He'll sit by the stream for three days straight," my dad told me. "Day and night. He just sits there, waiting. He can sit for three days without anything happening. And then suddenly, there it is. A four-pound eel!"

Patience was apparently the first prerequisite. You had to give the eel your time. We thought of it in terms of a transaction.

We also tried different kinds of bait. We put frozen shrimp on the hook. We tried plump slugs and beetles. Nothing worked much better than anything else. Once we found a dead frog in the grass by the stream. It was thick and shiny; we might have accidentally stepped on it. Dad put it on the hook and threw it in, but the next morning it was gone and the hook clean. So we went back to worms and kept working on our investment. One day, the big eel would come.

It never did, which only contributed to the eel's mystique. I think it was what made my dad an eel fisher. He was always telling me about glass eels, yellow eels, and silver eels, about how they changed shape, about eels older than any human, eels living in cramped, dark wells. He told me about their long journey across the Atlantic, back to their birthplace, a place far beyond anything I knew or could even imagine, about how they navigated using the movements of the moon, or maybe it was the sun, and about how every eel for some unfathomable reason simply knew where to go. How could they be so sure about something like that? How could anyone feel such overwhelming conviction about the path he or she had chosen?

When dad talked about the Sargasso Sea, it sounded like a magical fairy-tale world. Or like the end of the world. I pictured mile after mile of open sea that suddenly turned into a blanket of seaweed teeming with life and movement and eels writhing around one another and dying and sinking to the ocean floor while tiny see-through willow leaves floated up toward the light and let the invisible current take them. Every time we caught an eel, I looked into its eyes, trying to catch a glimpse of what it had seen. None of them ever met my gaze.

5

Sigmund Freud and the Eels of Trieste

How much can you ever really know about an eel? Or about a person? It turns out the two questions are related.

Sigmund Freud was nineteen when, in 1876, he picked up the gauntlet thrown by Aristotle more than two thousand years previously, which had been picked up in vain by others so many times before. He was the person destined to find the holy grail of natural science: the testicles of the eel.

Freud was born in 1856 in Freiberg in Moravia (now Příbor in the Czech Republic), but his family moved to Vienna before his fourth birthday. Even as a child, he was an excellent student, with an interest in literature and a remarkable talent for language; he enrolled at a university in Vienna when he was seventeen. Freud was primarily a medical student, but he also studied philosophy, physiology, and zoology under the renowned professor Carl Claus.

Claus specialized in marine zoology, was a fervent Darwinist and a leading expert in crustaceans, and like everyone in his field, he had an interest in eels. He had conducted research on hermaphroditic animals, of which the eel was still popularly believed to be one, and in addition to his professorship at the University of Vienna, he was also the head of a marine research station in Trieste.

During the first half of the nineteenth century, the eel question had lain dormant. Since Carlo Mondini had found and provided a plausible description of the reproductive organs of the female eel, it seemed it would simply be a matter of time before the male organs were found and identified also. And once they had been so located, the intractable mystery of the eels' procreation would be solved.

That being said, a lot of people were unconvinced by Mondini's discovery. One skeptic was the Italian scientist Lazzaro Spallanzani, who would eventually go down in history as the person who successfully dismissed spontaneous generation. Spallanzani traveled to Comacchio himself to investigate Mondini's findings and dismissed them as improbable. It was, of course, also a matter of prestige. So many prominent researchers had tried for so long to explain and describe the organs responsible for and the method of the eel's reproduction. Why had no one else succeeded? One single eel with reproductive organs and roe after all those years? Why couldn't any more be found? No, Mondini's eel seemed unique. It seemed implausible. And besides, sometimes, objective probability is less important than what people want to believe. In the scientific world, a lot of people simply didn't want to believe in Carlo Mondini's eel.

In Germany, the search for the eel's reproductive organs became, for a while, a popular spectacle. A reward of fifty marks was offered to any person who could find an eel carrying roe. Newspapers all over the country wrote about it. The eels were to be sent to a certain professor Rudolf Virchow, who would conduct a careful examination of each one; the German fishing authorities had agreed to pay the postage. The fanfare and the generous award resulted in a large number of eels being packaged and posted. Hundreds of eels from every part of Germany—half-eaten eels, rotting eels, eels crawling with parasites. The packages flowed in at such a rate that the fishing authority almost went bankrupt. And still, no sexually mature eel with roe was found.

It was only in 1824 that Martin Rathke, a German professor of anatomy, was able to find and adequately describe a female eel with fully developed reproductive organs, independent of Carlo Mondini. In 1850, Rathke also found an eel with fully developed eggs inside. It turned out Mondini had probably been right all along; his description of the reproductive organs tallied with Rathke's, but the eggs in Mondini's eel had been much smaller, as they were less fully developed.

With the first half of the biological equation verified, the hunt for the second part, the mythical testicles, could begin in earnest. But it was slow going at the outset. Many researchers still chose to believe that eels were hermaphroditic. The fatty tissue found adjacent to the reproductive organs in the mature females was in fact probably the male organs. How else could the answer to the mystery have eluded science for so long?

Laypeople by and large also preferred to cling to older, slightly more fanciful theories. In 1862, an amateur researcher,

David Cairncross, published a book entitled *The Origin of the Silver Eel*, in which he revived an old belief held by Sicilian fishermen that the eel's first manifestation was in fact a beetle, and that its past as an insect was proved by its ability to get by equally well on dry land and in water.

Almost one hundred years after Carlo Mondini's discovery, in 1874, a Polish zoologist, Szymon Syrski, announced that he and his colleagues at the natural historical museum in Trieste at last had found something that might be a mature male eel. Inside it, he had located a small, lobe-shaped organ that differed from the descriptions provided by Mondini and Rathke. It might, in fact, be the long-sought eel testicle. But since Syrski was unable to sufficiently describe the organ and prove it really did produce semen, nothing was certain. The scientific community required additional observations.

Thus in March 1876, Carl Claus decided to dispatch one of his young students from the University of Vienna to his research station in Trieste. And that is how at the age of nineteen, Sigmund Freud suddenly found himself in a simple laboratory on the Mediterranean with a knife in one hand and a dead eel in the other.

THE NINETEEN-YEAR-OLD SIGMUND FREUD WAS A YOUNG MAN WITH big plans. The year before, he'd visited Manchester and loved it, even the rain and the climate. He was keen to travel more and was, above all, eager to spend more time on practical scientific work, learning more about everything, making discoveries, describing things, understanding things. He loved the laboratory. What he saw through the microscope was always

unequivocally true; there was no room for prejudice or su-
perstition. All human knowledge came from the laboratory.
He envisioned a life in the service of science, possibly in England, maybe somewhere else entirely. And he was seriously
considering dedicating his life to natural science, to biology
or physiology, the tangible and concrete. In a family portrait
from 1876, he can be seen standing in the middle with his
hand on the chair of his mother, Amalia, the tallest of his siblings, wearing a three-piece suit, with his hair parted to the
side and a dark, well-trimmed beard. He's looking straight
into the camera, his gaze steady, as though nothing in the
world could perturb him.

It was this nineteen-year-old who in the spring of 1876
arrived in Trieste, with the ambition of solving the mystery
of the eel and leaving his mark on the history of science.
Trieste, located in the northeast corner of the Adriatic Sea,
belonged at this time to the Austro-Hungarian Empire and
was an important metropolis, home to a naval base and a
large port. Since the completion of the Suez Canal in 1867,
it had also been a gateway to Asia. Coffee, rice, and spices
were unloaded at the city's docks. Ships came from all over
the world, and people gathered there from all over Europe:
Italians, Austrians, Slovenes, Germans, and Greeks. As early
as Roman times, Trieste had been a meeting point and a site of
pilgrimage, a place where all kinds of languages and cultures
rubbed shoulders. Compared with Freiberg or Vienna, it was
almost certainly a city that made an impression, complex and
elusive.

So what did young Sigmund Freud find in Trieste? Quite
a bit is known about that, since he wrote several letters to his

childhood friend Eduard Silberstein describing his experience. He wrote in Spanish—since the two of them had become
close while studying that language—about the city, its restaurants, shops, and residents. On occasion, his word choices are
peculiar, possibly on account of Spanish not being his native
tongue, but more likely as a kind of code between friends.

In his first brief letter, from March 28, Freud writes that
Trieste is a very beautiful city and that "las bestias son muy
bellas bestias"; its beasts are very beautiful beasts. By *beasts*,
Freud meant women. During his first few days in Trieste, the
city's women seem to have fascinated him more than anything
else. In his letters, he writes about being struck during his
first day in town by the fact that every woman he met looked
like a "goddess." He describes their appearance and physical qualities in detail, saying they're tall and slim with long
noses and dark eyebrows, that they're paler than they should
be and have beautiful hairstyles and that some of them leave
a lock free to hang down in front of one of their eyes like a
tempting hook. He visits the neighboring city of Muggia and
writes about how the women there must be particularly fertile since virtually every other woman he saw was pregnant
and that the local midwives probably have no trouble finding work. He speculates ironically about whether the women
might be affected by "the marine fauna," making them "bear
fruit year-round," or whether they procreate at certain times
all together. "These questions will have to be answered by
future biologists."

He observes and describes the women almost like a scientist, but at the same time, they're alien to him, like members of
a different species. Freud does not, however, seem to have made

any close female acquaintances in Trieste, and before long, his mood and attitude toward the city changed. He starts expressing frustration with his situation in his letters to Silberstein: at the women who tempt and attract him, both younger and older ones, but who also confuse him emotionally. He remarks on their overuse of makeup. He writes about how they have a habit of sitting in their windows, looking out, smiling and meeting the eyes of men; he complains, slightly ironically, about having to distance himself from them, on account of his work.

Then, suddenly, he writes that all women in Trieste are "brutta, brutta," exceedingly ugly. It's as though he's uncomfortable with the realization that his feelings won't conform to the model of the cold, systematic man of science he strives to be. "Since we are not allowed to dissect people, I have nothing to do with them," he writes, after noting that in Trieste, even young girls use makeup.

As though to steel himself against the distraction of his sexual confusion, Freud instead focuses on his work. He has his own room at the laboratory, which is located a stone's throw from the Adriatic Sea. "I'm five seconds from the most recent Adriatic wave," he writes to Silberstein, and then gives a detailed description of his workplace:

> My little room has an odd floorplan, one window, in front of which is my worktable, with a great number of drawers and a large top, a second table for books and ancillary implements, three chairs, and several shelves holding some twenty test tubes. Last but not least, there is also a sizeable door, which, if you follow

*its lead, takes you outside. On the left side of the table,
in the corner, stands the microscope, in the right
corner the dissection dish, in the center four pencils
next to a sheet of paper (my drawings are therefore
cartoons, and not without value), in front stands a
series of glass vessels, pans, bowls, troughs containing
small beasts or bits of larger ones in seawater. In
between stand or lie test tubes, instruments, needles,
cover slips, microscope slides, so that when I am busy
working there is not a spot left on which I can rest my
hand. I sit at this table from eight to twelve and from
one to six, working quite diligently."*

Every morning, Freud goes to meet the fishermen as they
come into port with the catch of the day—baskets full of fat
Adriatic eels—then heads straight to the laboratory and sets
to work. He explains the object of his assignment to Silberstein, attaching simple drawings:

*You know the eel. For a long time, only females of
the species were known; even Aristotle didn't know
where the males came from and therefore claimed
that eels sprang from mud. Throughout the Middle
Ages and even in our modern times, there has been
a veritable frenzy to find a male eel. Within zoology,
where we don't have access to birth certificates and
where creatures—in accordance with Paneth's
ideals—act without first being observed, we cannot
say which is female and which is male unless the
animals display external differences. That there are*

*in fact differences between the sexes has to first be
proved, and only an anatomist can do so (since the eel
is incapable of keeping a diary from which we could
draw conclusions regarding its sex); he dissects them
and discovers either testicles or ovaries. . . . Recently,
a zoologist in Trieste claimed to have found testicles,
and thus to have discovered the male eel, but since
he apparently didn't know what a microscope is, he
failed to provide an exact description of them.*

Day in and day out, Freud sits by his desk in the laboratory, cutting up eels, searching, peering through his microscope and making notes, seeking the answer to the mystery. All answers are bound to appear underneath the microscope—that is the promise of science, and if you can't trust that, then what is there left to believe in?

But Freud doesn't find any eel testicles, and he gradually grows more frustrated. Every night at half past six, he takes a walk through the narrow alleyways of Trieste, past shops and restaurants, toward the sea, where the setting sun turns the water into a mirror, hiding all life underneath the surface; he hears dockworkers speaking German, Slovenian, and Italian, smells the spices and coffee, sees the fishermen pack up the last of their catch, sees the women with their made-up eyes moving toward the bars in the square. He sees all that . . . and thinks about eels.

*My hands are stained by the white and red blood of
the sea creatures, all I see when I close my eyes is the
shimmering dead tissue, which haunts my dreams,*

and all I can think about are the big questions, the
ones that go hand in hand with testicles and ovaries—
the universal, pivotal questions.

For close to a month, Freud sits in his simple laboratory, engrossed by his monotonous and fruitless work, but in the end, he has to admit he's failed. He hasn't been able to find what he came to seek: the reproductive organ of the male eel and the definitive answer to the eel question. "I've tormented myself and the eels in a vain attempt to discover the male eel, but all the eels I've dissected have turned out to belong to the fairer sex."

It was young Sigmund Freud's first scientific assignment, and failure was his fate. For weeks on end he stood by his desk, doggedly cutting up eels and searching their cold, lifeless bodies for reproductive organs. Long days, reeking of dead fish, covered in sticky eel slime. And not one testicle did he find. Freud examined over four hundred eels and none could be shown to be male. He knew exactly where in the eel to look, and he could describe what the organs ought to look like, but even so, he never found what he was looking for.

In one of his letters to Eduard Silberstein, Freud drew an eel swimming through the text. Its lips are curled in a faintly mocking smile. In the same letter, he spoke of the eels using the word he had previously used to denote a different, but equally enigmatic creature: "las bestias."

SO WHAT *DID* SIGMUND FREUD FIND IN TRIESTE? POSSIBLY, IF NOTH-ing else, an initial insight into how deeply some truths are hidden. In terms of both eels and people. And thus, the eel came to influence modern psychoanalysis.

Nineteen-year-old Freud was an ambitious young scientist. He'd gone to Trieste to write a groundbreaking report that answered, once and for all, the question that had confounded science for centuries: How do eels reproduce? He probably learned a great deal about the importance of patient and systematic observation in research, knowledge he would later apply to his patients on the therapy couch.

He'd also come to Trieste with an unshakable faith in science and in the rewards that await a person who's willing to work hard for them. But the eel forced him to confront his own, and science's, limitations. He found no truth under his microscope. The eel question remained unanswered. Completing his report a year later, he had to admit nothing could be proved about the sex and procreation of eels. He concluded with almost self-abnegating matter-of-factness: "My histological examination of the lobe-shaped organs will not permit me definitively to state the opinion that they are the testicles of the eel, nor does it give me substantial reason to reject it."

The eel eluded Sigmund Freud; perhaps that was one of the reasons he ultimately abandoned the pure natural sciences for the more complex and unquantifiable field of psychoanalysis. The way the eel eluded him was especially ironic, given what Freud would eventually focus on: it concealed its sexuality from him. The man who would come to define twentieth-century thinking about sex and sexuality, and who would delve deeper into the inner workings of humans than anyone before him, could not, where eels were concerned, even locate their sex organs. He had gone to Trieste to find an eel's testes but discovered only an enduring enigma. He wanted to understand the sexuality of a fish, but found, at best, his own.

It was also ironic because Freud's relationship with aquatic creatures was already slightly complicated. Much has been written about young Freud's relationship with a girl named Gisela Fluss. It began in 1871, when the then-fifteen-year-old Freud lived for a time as a lodger with Gisela's family in Freiberg. Freud was clearly attracted to Gisela, who was then only twelve, and expounded on how beautiful and alluring she was in letters to, among others, Eduard Silberstein. It may have been his initial sexual awakening, but, be that as it may, it ended in frustration and suppression. When Gisela married someone else a few years later, Freud gave her the moniker *Ichtyosaura*, or "fish lizard," after the scientific name for the prehistoric aquatic reptiles who were contemporaries of the dinosaurs.

To Freud, it was obviously a form of adolescent wordplay; *Fluss* means "river" or "flow." Gisela, as a member of the Fluss family, was a kind of sea monster, representing everything repressed and frustrating, such as sexuality, which moves furtively beneath the surface. That Freud chose a prehistoric water creature for her nickname was perhaps also his way of telling himself that the youthful and uncontrollable passion he'd felt for her would now belong to his past. He wouldn't let himself be seduced like that by anyone or anything ever again—until *las bestias* of Trieste appeared like the symbolic offspring of this his first *Ichtyosaura*.

After his stay in Trieste, it would be years before Sigmund Freud approached the subject of sexuality again, but once he did, it was hidden or repressed sexuality that interested him.

His theory about castration anxiety takes as its starting point the assumption that a child will at an early age develop a fear of being castrated, of being maimed and stripped of his or her sex, diminished and rendered harmless. Boys at the age of four or five are filled with unconscious sexual longing for their mother and feel in competition with their father. They perceive a threat, a fear of being punished for their urges, but they also feel shame and inferiority; this makes them realize their own insignificance in the world, which leads to the development of self; in due course, their yearning for their mother is replaced by identification with their father. And the pivotal moment in this process is, according to Freud, when a boy realizes women do not have penises. That is, he sees the woman, sees the *absence* of a male sex organ, and in that moment becomes aware of himself and his place in the world.

Freud's theory of penis envy is related to castration anxiety but deals with the psychosexual development of women. Girls are, like boys, at first closely bonded to their mothers, he claimed; it's when they first discover that they themselves have no penis that they slowly start dissociating from their mother and become drawn instead to their father. Girls see the penis as an attribute that symbolizes power and activity. Learning, in this way, their place in the world, they develop envy and experience guilt, which is projected onto their mothers. They can see what they lack, see the *absence* of a male sex organ, and in that moment become aware of themselves and their limitations.

These theories have been challenged many times since they were first formulated, and from many different perspectives. Can the male sex organ, or the possession or lack of it, be such a pivotal detail in the psychosexual development of humans? It seems absurd and a little ridiculous. These are theories from a different time, which grew out of a different historical context. They are also theories that dodge the accepted scientific method. They operate within the suppressed and the concealed. They can't be systematically observed or verified or rejected. They are not the kinds of truths a microscope can reveal.

And yet they must be founded on the basis of some kind of experience. We can picture the young scientist in a cramped laboratory in Trieste. He is far from home in a strange city, and he is wearing a white coat and glasses, with a well-trimmed, dark beard. He is standing by a desk in front of a small window, with a sticky dead eel in his hand. And he's looking through his microscope, as he's done four hundred times before, and what he can see through the lens is no longer just an eel, it is also himself.

DESPITE THE CONCERTED EFFORTS OF THE YOUNG FREUD, THE MYStery of the eel's reproduction remained unsolved for a while longer. In 1879, a German marine biologist, Leopold Jacoby, wrote, somewhat dejectedly, in a report for the US Commission of Fish and Fisheries:

"To a person not acquainted with the circumstances of the case, it must seem astonishing, and it is certainly somewhat humiliating to men of science, that a fish which is commoner in many parts of the world than any other fish . . . which is daily seen at the market and on the table, has been able in spite of the powerful aid of modern science, to shroud the manner of its propagation, its birth, and its death in darkness, which even to the present day has not been dispelled. There has been an eel question ever since the existence of natural science."

What neither Freud nor Jacoby knew was, of course, that eels have no visible sex organs until they need them. Its metamorphoses are not just superficial adaptations to new life conditions. They're existential. An eel becomes what it needs to be when the time is right.

Twenty years after Freud's failed efforts, a sexually mature male silver eel was finally found off the coast of Messina in Sicily. And thus, the eel had finally become a fish. A creature not so dissimilar from others.

6

Illegal Fishing

At times, we fished illegally. It was above all a matter of convenience. Because while the narrow path might be the right one, sometimes the wide one is so much easier to walk. Since Nana and Grandad's fields bordered the stream, we were allowed to fish in it, but only on our side, the farm side. Which was also the difficult side, with the tall grass and the steep, muddy banks. On the other side of the stream, everything was different; there, a flat meadow stretched all the way to the water's edge. The fishing rights on that side were owned by the fishing club in town.

The other side of the stream was the stuff of dreams. Not only because it looked so accessible, but also because it symbolized something we perceived as unjust. On the weekends, the members of the fishing club would stand there on the flat ground in their green sport jackets with multiple pockets, their expensive fly fishing rods and ridiculous little hats, swinging their shiny, thick lines over their heads to try to

catch one of the rare salmon that constituted the upper ech-
elon of the stream's class hierarchy.

We'd never once seen salmon in the stream. At least not
live salmon. Dad found an enormous dead salmon once. It was
floating belly-up; he brought it home. It was fat and bloated
and weighed more than twenty pounds. It also smelled pretty
bad. We buried it, after admiring it with our hands over our
mouths and noses.

One summer, Dad acquired an old wooden rowboat. He
saw it advertised in the paper and bought it for two hundred
kronor; we sanded and painted it out on the lawn. It was
moored to the willow tree just above the rapids, and one night
when we reached the stream, he suggested we row across and
set up our spillers on the other side instead. The thought
had never even crossed my mind, but suddenly it seemed
completely rational. There was, for obvious reasons, no one
on the other side at that time. And besides, it was the same
stream; the difference between fishing here and fishing there
was entirely theoretical. Moreover, how could anyone claim
to hold the rights to something as transient as flowing water?

"But if the train comes, we'll have to hide," Dad cau-
tioned. The railway ran along an embankment next to the flat
meadow. It came around a bend a few hundred yards from
where we were and then ran parallel to the stream, with an
unobstructed view of the meadow all the way down to the water's edge. And maybe there would be a member of the fishing
club on it this particular night who would see us poaching
and sound the alarm, catching us red-handed like the crimi-
nals we were.

We rowed across and moored the boat; I was both terrified and exhilarated. Then we picked up our things and walked along the stream, commenting on just how much more convenient this side was. It wasn't merely the stuff of dreams, it was real, and there was no tall, wet grass to slog through and no muddy banks to slide down. I told myself it was virtually our moral obligation to fish there.

But we set up our spillers faster than usual, glancing nervously at the railway all the while, poised to flee at the first distant sound of the approaching train. When it did come, it careered through the bend so much faster than I could've imagined; we turned off our flashlight and threw ourselves down in the grass. I pressed myself against the ground, doing my best to disappear among the tussocks, hiding my face and holding my breath. The train thundered past and the whole meadow was illuminated like when lightning makes time stop and I imagined we really were invisible and that my dad was lying there just like me, with his hands over his face, not breathing.

Now I'm thinking he was probably smiling. That he wasn't scared of being caught at all—Why would anyone care? How would they identify us?—but was playing along for my benefit. That he staged the whole spectacle to make it more exciting. Maybe he was worried I would grow tired of it all otherwise.

I don't know why he would have been worried about that—there was nothing I liked more—but it's also only now, much later, that I've started to wonder if Dad really ever went eel fishing as a child. I'd always figured he must have. I'd always thought he and I were carrying on a tradition that had begun long before either of us. That he was doing for me what someone else had done for him and that those nights down by the river constituted some kind of continuity across time and generations. Almost like a ritual.

But he certainly never fished with his father (the man he called Father). My grandfather (the one I called Grandfather) didn't fish. He never did anything that wasn't immediately useful. He worked and he rested and when he ate he did it quickly and in silence. He was a teetotaler and hated the effects of alcohol; as far as I knew, he had never in his life taken so much as one day off, had never traveled anywhere, never been abroad. Wasting time and energy on something as seemingly frivolous as eel fishing was not for him. It had nothing to do with patience, it was more a matter of obligation. The narrow path looks different to different people.

Maybe Dad fished alone, or with someone else entirely, but if so, I don't know anything about it. I remember Dad telling me how much fish there used to be in the stream a long time ago, about how the bottom crawled with eels and how the surface turned silver when the salmon traveled up it in the spring. But he didn't speak from experience; these were stories from before he was born that he'd picked up somewhere. His own stories about caught or lost eels I knew already, because I'd been there with him. His stories were my stories. It was as though there had been nothing before us.

Was that the case? Did it start with the two of us? If so, did it have anything to do with the fact that the person he called Father and I Grandfather was really someone else? Were our nights by the stream an attempt to compensate for something my dad hadn't had, to realize his own vision of what a father and son could be to each other? A way of forging his own narrow path through life?

7

The Dane Who Found the Eel's Breeding Ground

How far do you have to be prepared to go to understand an eel? Or a person? Johannes Schmidt was twenty-seven years old when he stepped aboard the steamship *Thor* in 1904 and set off to find the birthplace of the eel. It would be almost twenty years before he reached his destination. A few years after he did, a British marine biologist, Walter Garstang, would write an ode to Schmidt, which was eventually published in what might very well be the only collection of poems ever written about the larval stage of various animals, *Larval Forms, with Other Zoological Verses*.

> *All honour to the Danes who solved*
> *This age-old mystery,*
> *Who, step by step, and year by year,*
> *Revealed the history:*
> *Johannes Schmidt the leader,*

With "Papa" Petersen behind,
Who made the "Thor" and "Dana" ships of fame
For all mankind

Quite a lot had happened in humankind's dogged quest to understand the eel's life and existence since Sigmund Freud's fruitless search for testes in Trieste. A Danish marine biologist, C. G. Petersen, had in the 1890s managed to observe the last metamorphosis of the eel and proposed that all eels reproduce in the sea. Even Aristotle had, as we know, noticed that fully grown eels sometimes move out into the sea, and in the seventeenth century, Francesco Redi had noted that glass eels appear along the coasts in the spring and wander up rivers. But Petersen was able to describe how it happens in more detail. In particular, he successfully observed and described how yellow eels turn into silver eels. Until then, a lot of people were unconvinced that the two belonged to the same species. Petersen demonstrated unambiguously that they were each manifestations of the same fish. He saw the silver eel's digestive organs shrink and saw it stop eating, saw its reproductive organs develop and its fins and eyes change. The transformation was apparently the eel's way of preparing for procreation.

In 1896, two Italian researchers, Giovanni Battista Grassi and his student Salvatore Calandruccio, had also explained the first metamorphosis of the eel. They had made a comparative anatomical study of different kinds of larvae caught in the Mediterranean to glass eels, and drew from it the conclusion that a small creature shaped like a willow leaf called *Leptocephalus brevirostris* must be the very first form of the European eel, *Anguilla anguilla*. This larva had previously been believed to be its own species. Now it became clear it was in fact an eel. What's more, Grassi and Calandruccio were also the first people ever to witness the metamorphosis, when a small willow leaf in their aquarium in Messina on Sicily miraculously turned itself into a glass eel.

It was a sensational discovery. "When I reflect that this mystery has occupied the attention of naturalists since the days of Aristotle, it seems to me that a short extract of my work is perhaps not unworthy to be presented to the Royal Society of London," Grassi wrote in a report that would eventually be published in what was at the time one of the world's most prestigious scientific journals, *Proceedings of the Royal Society of London*. In his report, Grassi also noted that this particular kind of larva, which had now been shown to be the first incarnation of the eel, had relatively large eyes and consequently was probably hatched at great depths. Possibly, he proposed, in the Mediterranean.

By the early twentieth century, it was known, then, that the yellow eel turns into the sexually mature silver eel and wanders back into the sea in the autumn, never to return. It was also known that leptocephalus larvae turn into tiny, delicious glass eels that appear around the coasts of Europe in spring in search of a place where they can live and transform themselves into fully grown yellow eels. But what happens in between? And where does it happen?

When the German zoologist Carl H. Eigenmann gave a speech to the American Microscopical Society in Denver, Colorado, in 1901, he entitled his lecture "The Solution of the Eel Question." It was not intended literally. He was still unable to provide the ultimate solution to the eel question. On the contrary, he cited a scientific anecdote according to which "all important questions have now been answered, save the eel question." But, Eigenmann explained, the question itself had changed. Before, the eel question had been about what the eel truly was, a fish or something else altogether. It had been about the eel's propagation—about finding its reproductive organs, about whether the eel gave birth to live young, whether it was hermaphroditic or not—and about what its metamorphoses signified.

But now, at the dawn of the new century, the eel question was this: What do mature eels do after moving back into the sea? When and where do they breed? And where do they die?

SO WHERE DID THE SILVER EELS GO? AND WHERE DID ALL THE MYS-terious willow leaves come from? Where was the eel's birthplace? That was what the twenty-seven-year-old Johannes Schmidt set out to find in the spring of 1904.

Johannes Schmidt was a marine biologist from Denmark. He lived his first few years in a small redbrick house on the grounds of Jægerspris Castle on North Zealand, about thirty miles north of Copenhagen, where his father was the steward. He was brought up in a warm and sheltered environment, surrounded by woods and nature, far from the big city and the world of science, and even farther from the Sargasso Sea.

At the tender age of seven, however, Johannes Schmidt lost his father, and he, his mother, and his two younger brothers were suddenly forced to move to Vesterbrogade in Copenhagen, one of the city's liveliest streets, and to a very different kind of life, surrounded by different kinds of people. It was an upheaval that affected Johannes Schmidt's life not only emotionally but also practically. The Carlsberg brewery was located just a few hundred yards from his new home, and even closer was the home of Johannes Schmidt's uncle, Johan Kjeldahl, who worked as a chemist at Carlsberg's research laboratory, where Schmidt would eventually begin his own scientific career.

The same year the seven-year-old Johannes Schmidt moved to Copenhagen with his family, the world-famous chemist Louis Pasteur visited the city. Pasteur had developed a method for protecting food from bacteria and microorganisms; pasteurization, as it had been named in his honor, had been hugely significant for beer breweries. When Pasteur came to Copenhagen, he was consequently invited to visit Carlsberg, and a proud J. C. Jacobsen, the brewery's owner, was so impressed by the great scientist he decided to invest in a sophisticated in-house research laboratory. In addition to brewing beer, Carlsberg would also pursue modern, advanced research. And not just about beer making and food conservation but groundbreaking basic biological and natural scientific research. It was a matter of prestige but also a commercial calculation. Over time, it helped Carlsberg grow from a small family-owned brewery to one of the world's largest, while the company's research department would also, in roundabout, indirect ways, contribute to bringing the eel and humankind a little bit closer together.

After moving to Copenhagen, during his first years at school, Johannes Schmidt began to spend more and more time in the Carlsberg research laboratory, shadowing his uncle Johan Kjeldahl, with whom he also lived for a time. It was there, in the laboratory, that he learned the basics of scientific work. It was also where a passion for science—that compelling need to observe, describe, and understand—was awakened in him. When he eventually embarked on his successful academic career, and traveled the world in pursuit of his research, it was with the financial support of Carlsberg.

Johannes Schmidt received a degree in botany and a grant to study the vegetation of what was then known as Siam (now Thailand) in 1898. In 1903, he submitted a doctoral thesis on mangroves, only to immediately switch his focus to marine animals.

On September 17, 1903, he married Ingeborg van der Aa Kühle, whom he'd known since he first came to Copenhagen at the age of seven and who was the daughter of Søren Anton van der Aa Kühle, the successor to J. C. Jacobsen as director of Carlsberg. The wedding took place in Carlsberg's own church, the Jesuskirken in Copenhagen, and in the spring of 1904, the couple acquired an apartment of their own on Østerbrogade. They had barely moved their furniture in before Johannes Schmidt set sail to find the origin of the eel.

"THE PROBLEM OF THE PROPAGATION AND BREEDING PLACES OF the Common or Fresh-water Eel is one of great antiquity," Johannes Schmidt would later write in a report to the Royal Society of London. "From the days of Aristotle naturalists have occupied themselves therewith, and in certain regions of Europe it has exercised popular imagination to a remarkable degree."

He wrote *places*, in the plural, because how could anyone

know for certain there was just one breeding place? And he lingered on that enticing enigma, the one that had for centuries occupied so many scientists and that had now apparently ensnared him as well.

"We know, then, that the old eels vanish from our ken into the sea, and that the sea sends us in return innumerable hosts of elvers. But whither have they wandered, these old eels, and whence have the elvers come? And what are the still younger stages like, which precede the 'elver' stage in the development of the eel? It is such problems as these that constitute the 'Eel Question.'"

More specifically, there was one aspect of the eel question that bothered Johannes Schmidt. His Italian predecessors Grassi and Calandruccio had proposed that the eel, or at least the Italian eel, reproduces in the Mediterranean, since that was the only place they had found leptocephalus larvae. But at the same time, the larvae caught in the Mediterranean were large, three to four inches long, and clearly not newly hatched. How come no one had ever found smaller specimens?

As early as May 1904, mostly through sheer happenstance and before his mission had technically been made official, Johannes Schmidt managed to catch a leptocephalus larva in the sea just west of the Faroe Islands. It, too, was large, three inches long, but it was the first time anyone had seen an eel larva outside the Mediterranean, and it convinced Schmidt that Grassi and Calandruccio were likely mistaken about the eel's breeding ground. Schmidt also realized that in order to solve the mystery, he would have to trace the eel back to its source, looking for ever-smaller larvae, until somewhere in the vast ocean, he found the first newly hatched willow leaf, and thus the birthplace of the eel. He needed to find a needle in a haystack. And the haystack was an ocean.

"I had little idea, at the time, of the extraordinary difficulties which the task was to present, both in regard to procuring the most necessary observations and in respect of their interpretation," Schmidt would later write. That was, by all accounts, a polite and conservative understatement.

Between the years 1904 and 1911, Johannes Schmidt patiently sailed up and down the coasts of Europe with a trawl: through the waters off Iceland and the Faroe Islands in the north, across the North Sea off Norway and Denmark, south along the Atlantic coast of the continent, past Morocco and the Canary Islands, and into the Mediterranean, all the way to the Egyptian coast. He found lots of leptocephalus larvae, but they were all more or less the same size as the first one he'd caught, between two and half and three and a half inches.

After more than seven years of searching, he was still stuck on square one, and evidently plagued by a certain degree of despondency. "The task was found to grow in extent, year by year, to a degree we had never dreamed of," he wrote. "And this work has been handicapped throughout by lack of suitable vessels and equipment, and by shortage of funds; indeed, had it not been for the private support afforded from numerous different sources, we should have had to long since relinquish the task."

He had at least felt able to draw one firm conclusion: since all the larvae he'd found along the coasts of Europe were relatively large and evidently not newly hatched, he'd realized that eels probably do not reproduce near the coast and that his search would have to continue considerably farther out to sea. For this, the steamship *Thor* was insufficient; instead, Johannes Schmidt was able to enlist the aid of Danish shipping companies that sailed the Atlantic. He equipped their ships with nets and instructions, and between 1911 and 1914, twenty-three large freighters participated in the search for the tiny, transparent larvae. Their crews had no scientific training and no equipment other than the trawling nets Schmidt had given them, but they were under instruction to drag the nets along behind them, mark where they raised them and send their catches to the laboratory in Denmark. More than five hundred catches were logged by the freight ships covering large swathes of the northern part of the Atlantic.

Schmidt for his part set off in the summer of 1913 on the schooner *Margrethe*, which a Danish company had lent him. He scoured the waters all the way from the Faroes to the Azores, west toward Newfoundland and then south in the direction of the Caribbean.

The intensified search yielded results. Before long, Johannes Schmidt found that the eel larvae became more numerous as he moved west, while their size decreased. At one point, about halfway across the Atlantic, between Florida and West Africa, he caught a larva measuring only 1.3 inches, a new record. Eventually, pushing even farther west, he found a specimen measuring less than 0.7 inches.

Schmidt collected all the fragile leptocephalus larvae, from both his own expeditions and those of his helpers, studied them under a microscope, measured them, and kept meticulous notes: length and number, depth and date, latitude and longitude. Slowly but surely, he built up an enormous collection of data, which guided him, almost imperceptibly slowly, toward his goal. Among other things, he was able to discern a link between the tiny willow leaves' movements across the Atlantic and the mighty ocean currents. He also found something else, almost by chance.

It was already known that the eels that swim up rivers and other waterways on the American continent belong to a different species from their European counterparts. The two types of eel are virtually identical, and they go through the same metamorphoses, but they nevertheless belong to different species of the *Anguilla* family. The only thing that differentiates them is that the European eel, *Anguilla anguilla*, has a few more vertebrae then the American eel, *Anguilla rostrata*.

Johannes Schmidt's mission was, of course, to find the birthplace of the European eel, but what he discovered as he pushed farther and farther west was that more and more of the larvae caught belonged to the American species. That posed certain problems. Aside from measuring and counting the larvae, he now also had to classify each specimen. Out on the ocean, aboard a rolling and pitching ship, he had to place every last tiny willow leaf under the microscope and try to count the muscle fibers along its back; the fibers correspond to the number of vertebrae that appear in the fully grown eel. By so doing, he could determine which species the larva belonged to, and then construct tables showing where each species was more common. What he discovered was that in the western part of the Atlantic, the population was mixed. European and American larvae comingled, seemingly powerless to resist the currents, and they were caught in the same nets. That should, logically speaking, have meant that European and American eels were not only virtually identical but that they also bred in the same spot.

If that were the case—and in turn meant that Schmidt, if he could find the birthplace of the European eel, would by default also find the birthplace of the American eel—only one mystery would still remain: How do they know which species they are? How do the tiny willow leaves drifting through the Atlantic know where to go? Clearly, Schmidt wrote, larvae of both eel species travel together on the Gulf Stream, but at some point, their journeys diverge; the American larvae suddenly veer west, turn into glass eels, and wander up American waterways, while the European ones press on eastward. "How," Johannes Schmidt wrote, "do the masses of larvae in the western Atlantic sort themselves out, so that those individuals which belong to *Anguilla anguilla* ultimately find themselves in Europe, while those of *Anguilla rostrate* 'land' on the shores of America and the West Indies?"

His conclusion was that the different types of larvae, similar as they may appear, are programmed from birth to seek out different destinations. Simply put, the American ones grow faster than their European cousins, meaning they have the strength to break out of the mighty ocean current as it passes the American coast, instead of drifting on toward Europe. American eel larvae undergo their first metamorphosis, turning into glass eels, after just one year, while the European ones spend two long years drifting with the currents, and become glass eels only after three.

This is what makes the eel unique, Johannes Schmidt argued. Not its metamorphoses, not that the mature silver eels wander back to the sea and cross an entire ocean to breed.

"The point which makes our eel an exception among fishes, and among all other animals, is the enormous extent of its journeyings in the larval stage."

THE SPRING OF 1914 FOUND JOHANNES SCHMIDT WITHIN TOUCHING distance of his goal. He was slowly homing in on the birthplace of the eel; all his observations were pointing in the same direction; all that was needed now were more expeditions. The scientific approach—empirical, systematic observation— had, after ten years of at times seemingly hopeless searching, come through after all. The truth would soon reveal itself in Johannes Schmidt's microscope. In May 1914, he found two eel larvae that were just a third of an inch long.

That was when more worldly affairs suddenly got in the way. First, the schooner *Margrethe* sank after running aground off the island of Saint Thomas in the Caribbean. Fortunately, the collected specimens could be salvaged, but, Schmidt wrote, "here we were, on Saint Thomas with no ship. The only thing to be done for the moment was to endeavor to press forward the work being done from the trading vessels."

Soon after that, in July 1914, the First World War broke out. Suddenly, the Atlantic was no longer just the enigmatic location of the eel's propagation but also a war zone. Submarines patrolled the sea, threatening any and all who dared to venture out; several of the trading ships participating in Schmidt's search were sunk; sailing around the ocean in search of transparent little willow leaves was no longer just a fairly unpromising endeavor, it was also deeply dangerous.

For five long years, Johannes Schmidt sat in his chamber,

waiting for the world powers' irrelevant skirmish to end so he could once more resume his much more urgent task. While he waited, he worked on the data he had already collected, photographed his specimens, cataloged them, drew up tables and diagrams. He was impatient, knowing exactly what he had to do "as soon as the war ceased."

In 1920, when large parts of Europe still lay in ruin, Johannes Schmidt set sail again. During the imposed hiatus, he'd made sure he would be even better equipped than before. Through the East Asiatic Company in Copenhagen, he had gained access to the four-masted schooner *Dana* and outfitted her with all the necessary scientific equipment. Most important, however, he now knew where to look.

During 1920 and 1921, the *Dana* caught more than six thousand leptocephalus larvae in the western part of the Atlantic. Schmidt was able to make a detailed map of where the smallest specimens had been found. Specimens so minute, Johannes Schmidt wrote, "that there can be no question . . . where the eggs were spawned."

A PERSON SEEKING THE ORIGIN OF SOMETHING IS ALSO SEEKING HIS own origin. Is that a reasonable statement? Was that true of Johannes Schmidt? The man who since the tender age of seven had lived with only fading memories of his father? Had he fished for eels as a child? Had he held an eel and tried to look into its eyes? In 1901, just a few years before he set off on his first journey, his uncle Johan Kjeldahl, who had at times been a kind of surrogate father, drowned. In 1906, while he was still sailing up and down the coasts of Europe, his mother passed away. The Johannes Schmidt who sailed west, out into the open ocean toward the unknown, was a young man whose every connection with his own origin had been severed.

What that really meant to him, we can't say for certain. There is in his background, or at least what we know of it, very little to explain why he spent his life seeking the eel's birthplace. Granted, he was a consummate scientist. He was often described as exceedingly efficient: he observed, described, and tried to understand; only rarely did he seem to trouble himself with the question of why. He took a matter-of-fact view of the world and his own place in it. In letters and reports, he was plainspoken and formal. In pictures, he looks warm and friendly and usually wears a three-piece suit and bow tie. He was said to love animals, with a particular love of dogs. But his motivation remains a well-buried secret. He grew up in a safe middle-class environment and felt at home in the world of science from an early age. By marrying Ingeborg, he also became a member of the upper echelons of Copenhagen's bourgeoisie. He could have chosen an easier, more comfortable life. In terms of common measures of success—wealth, prosperity, status—he clearly had more to lose than gain from his journeys. And yet it never seems to have occurred to him to question the usefulness of spending almost two decades drifting around the vast Atlantic Ocean, finding tiny transparent willow leaves.

Put plainly, Johannes Schmidt was entranced by the eel question, by the enduring mystery of where the European eel breeds, how it is born and how it dies. "I think," he wrote, "the eel's life-history is, in point of interest, hardly surpassed by that of any other species in the Animal Kingdom."

Perhaps there are people who simply don't give up once they've set their minds to answering a question that arouses their curiosity, who forge ahead until they find what they seek, no matter how long it takes, how alone they are, or how hopeless things seem. Like a Jason aboard the *Argo*, seeking the Golden Fleece.

Or perhaps the eel question provokes a different kind of doggedness among those who tackle it. The more I myself learn about the eel, and the more aware I become of what the acquisition of that knowledge has cost throughout history, the more I'm inclined to believe that is the case. Above all, I want to believe that the mystery draws us in because some aspect of it is familiar. The origin of the eel and its long journey are, despite their strangeness, things we might relate to, even recognize: its protracted drifting on the ocean currents in an effort to leave home, and its even longer and more difficult way back—the things we are prepared to go through to return home.

The Sargasso Sea is the end of the world, but it's also the beginning of everything. That's the big reveal. Even the pale yellow eels my dad and I used to pull out of the stream on late August nights had once been willow leaves, drifting four thousand miles from a strange and fairy-tale-like place far beyond what I could imagine. When I held them in my hands and tried to look into their eyes, I was close to something that transcended the limits of the known universe. That is how the eel question draws you in. The eels' mystique becomes an echo of the questions all people carry within them: Who am I? Where did I come from? Where am I going?

Was it like that for Johannes Schmidt?

Perhaps, but it is of course perfectly possible that all those things were completely inconsequential to him. He had accepted a challenge and decided to see it through. He had formulated his own explicit question—Where are eels born?—and a method that generated its own momentum, so to speak. He caught tiny transparent willow leaves, and with each specimen caught, the task became catching one that was smaller still. His goalposts kept moving. It was as simple as that.

And the eels, for their part, were there beneath his feet while he crossed the Atlantic, as they had always been. The tiny willow leaves drifting on the ocean currents in one direction and the fat, fully matured silver eels, their course stubbornly set for the Sargasso Sea, swimming in the other. Year after year, they continued their mysterious journey away from home and back again, unperturbed by world wars and human curiosity. Just as they had long before Johannes Schmidt ever set sail, long before Aristotle saw his first-ever eel and tried to understand it, long before the first human ever set foot on this planet. The eels didn't care about the eel question, and why would they? To them, it was never a question in the first place.

IN HIS EXHAUSTIVE REPORT, WHICH APPEARED IN *PHILOSOPHICAL Transactions of the Royal Society of London*, eventually published in 1923, Johannes Schmidt accounted for close to two decades of work. On a map, he demarcated the area he could with a considerable degree of certainty claim was the spawning site of the eel. The elliptical area almost exactly delineates what we today call the Sargasso Sea.

"During the autumn months," he wrote as a kind of summary, "the silver eels leave the lakes and rivers and move out into the sea. Once beyond freshwater limits, the eels are, in most parts of Europe, outside our range of observation. No longer subject to pursuit by man, hosts of eels from the most distant corners of our continent can now shape their course south-west across the ocean, as their ancestors for unnumbered generations have done before them. How long the journey lasts we cannot say, but we know now the destination sought: a certain area situated in the western Atlantic, N.E. and N. of the West Indies. Here lie the breeding grounds of the eel."

This is why we now know—at least with some degree of certainty—where the eel reproduces. All our knowledge on this matter rests on Johannes Schmidt's work. What we don't know is why. Why there? What's the point of the long, hopeless journey and all the trials and metamorphoses? What is there for the eel in the Sargasso Sea?

Johannes Schmidt might have replied that it's irrelevant. Existence comes first. The world is an absurd place full of contradictions and existential confusion; only those who have a goal are ultimately able to find meaning. One must imagine the eels happy.

And Johannes Schmidt as well. In 1930, he was awarded the prestigious Darwin Medal by the Royal Society of London. And, with that, he was done, his story complete. Three years later, he died from the flu.

8

Swimming against the Current

July and August was prime eel fishing season. Never before midsummer. "There's no point fishing before midsummer," Dad would say. "It's too bright, the eels won't bite, it has to be darker."

He talked a lot about eel darkness, when the nights are at their murkiest and the eels at their boldest, when out of a thirst for adventure or recklessness they expose themselves to humans.

But of course, he had it wrong. Or maybe he chose to believe his own truth because it made life a little easier.

There really is such a thing as eel darkness; it happens at the end of summer and lasts for a few months. This is when the silver eels begin their journey toward the Sargasso Sea and can therefore be lured into fishermen's traps along the coasts. Our eel darkness was something else. It happened when Dad was on summer leave and therefore able to spend his nights down by the stream instead of in bed.

He'd worked all his life. For as long as I'd been alive, and before that, too, he'd been a road paver. He got up every morning before six, drank coffee and ate sandwiches, and was at work before seven. He was part of a work team who—in relative freedom, a chain gang without chains—traveled around paving, making new roads or fixing old ones. It was heavy work, hot and foul smelling; someone got to drive the big machine that spread the asphalt out over the prepped road surface, but someone had to walk behind it, too, with a shovel or a rake, in a cloud of tar and soot. They worked on commission, so each step taken and each shovel lifted was a krona earned. They worked from seven to lunch, coffee and sandwiches in the work shed, then from lunch until four, unless there was an unusual amount of work to be done and they had to stay late.

He usually came home around half past four, took off his dirty work clothes and went straight to bed. His body was hot and sweaty, his whole being exhausted. You were allowed in his room, but he didn't say much. "Just need a bit of a rest." Sometimes he dozed off, but thirty minutes later, he'd get back up for supper and what was left of the day.

Work was more than an occupation, it was an integral part of him; it broke him down, but it also made him hardy, it shaped and colored him. He was a fairly large man, not too tall, but muscular and top heavy. He was tenacious and strong. His upper arms were powerful and firm; both my hands together weren't enough to encircle them. In the summers, he worked bare chested and got so tanned his skin looked like dark rust and the faded tattoo on his forearm, a simple anchor, grew almost invisible. (He'd gotten the tattoo before he was of age, drunk and lost in Nyhavn in Copenhagen, and why he'd picked an anchor was probably a mystery even to him, since he'd never been to sea.) His hands were big and ponderous with thick, leathery skin. One of his pinkies was missing; it had been broken so many times it had stiffened into a crooked grimace, like an oversized claw. He'd asked a doctor to remove it, and the doctor had obliged.

He'd worked for decades, and it showed. The warm, newly made asphalt he carried, shoveled, and flattened every day seemed to have seeped into his skin. He smelled profoundly of tar, even after washing up and changing his clothes. It was a mark of the working class.

When we were out driving, he would point to a paved street and say "I made that." He liked his work and could almost, if pressed, admit he was good at it. His professional pride was of the natural, universal kind—the kind that comes from knowing you're pretty good at something not a lot of people know how to do, and from knowing there's a certain permanence to what you make and that other people value it.

But his identity didn't revolve around being a paver. His profession was just a word. When he talked about himself, he called himself a worker, and contained in that concept were most of the things he considered central to his being. Nor did it seem to have been a matter of choice. He was a worker from birth and his identity was inherited. He was a worker because something bigger and stronger than him had chosen that life for him. The course of his life was predetermined.

But if that was his heritage, what was mine? Maybe—and herein lies the minute, barely perceptible shift that takes place between generations—a never-spoken but ever-present encouragement: No, all doors are not open to you, and time is shorter than you think, but, of course, you're always free to try.

DURING THE SUMMER VACATION, WE SOMETIMES WENT DOWN TO the stream earlier in the day, while there was still light. Instead of bats, swallows swooped and dived above the water; from a distance they looked almost identical but they moved differently. The sun glittered in the stream and the tall grass waved drily in the breeze.

One early evening, we were standing by the willow tree a distance below the rapids.

"Think you can swim across here?" Dad asked.

"Of course I can."

"I'll give you ten kronor if you cut straight across."

"Sure."

"But it has to be straight across. Straight across the current. You can't drift. If you swim straight across without drifting, I'll give you a tenner."

I undressed and stepped into the water. It was cold and dirty; I hesitated for a second or two.

"There," Dad said, pointing. "Straight across right here, from the tree to the rock on the other side."

I slipped down and out into the stream and started swimming; for about five feet, I did okay. I held my head up high and kept an eye on my target. Straight across to the rock. It didn't feel particularly insurmountable. But then I reached the middle of the stream where the current was at its strongest; it grabbed me like a hand brushing crumbs off a table. I was swept along a few feet, pulled under, I swallowed water and coughed before I managed to turn against the flow and stay motionless in the middle of the stream for a few seconds, like a boat at anchor, paddling frantically against the current. Suddenly, I felt it lift me up and shove me forward; I virtually hurled myself toward the shore. I climbed out on trembling legs, about fifteen feet downstream from the rock.

Dad laughed and pointed from the other side.

"You get one more chance. Since you have to come back, too."

"Can't you come get me with the boat?" I yelled.

"Oh, no. Come on. Straight across."

I walked over to the rock, shook the lactic acid out of my muscles and stepped back into the water. This time, I aimed upstream from the outset, launching myself out; the momentum helped me swim diagonally against the current for a brief moment. For those few seconds, I was also on the right side of the willow tree on the other side, but then the water caught on to what was happening and wrestled me violently downstream. I managed to steer my way to shore, grab a branch, and pull myself up onto dry land, three feet or so from the willow tree.

"That's close, who would've thought?" Dad said and turned around to go get our fishing gear.

I stayed where I was, letting the last rays of the setting sun dry me. When he came back, I got dressed and we walked silently along the stream, out onto a narrow spit of land where we fished while we waited for it to be time to set up our spillers.

I caught a small perch, which had swallowed the hook so badly we had to break its neck; Dad said we could try using it for bait. As the sun winked out below the horizon, a bat flew quickly and silently over our heads.

"I guess it's time," Dad said. I never did get that tenner, of course.

9

The People Who Fish for Eel

Hanö Bay on the east coast of Skåne in Sweden is home to a unique beachfront that stretches for about thirty miles, between Stenshuvud in the south to Åhus in the north. This is what's often called Sweden's eel coast.

It's a pretty landscape, but not in a pastoral or exaggerated way. There is natural beauty there, if of the somewhat inaccessible kind. Hanö Bay's coast is gently rounded, wreathed by a sparse, windswept pine forest. A long, narrow, almost white beach, often visible from the road, lines the edge of the forest on the seaward side. It looks like a discarded, sun-bleached strip of fabric running the length of the bay. The sea is shallow and the water a deep shade of blue.

Big, thick wooden posts rise out of the sand at regular intervals, seven or eight in each small cluster. They look like telephone poles, but without wires, erected seemingly at random. These poles were used to hang up fishing equipment and nets, to dry and mend them, and wherever you see a cluster of poles sticking up at the horizon, you can be almost sure you will also find a small house. Usually an old brick or stone building, often with a thatched roof, sometimes half-buried in the dunes, almost always facing the sea. These houses are called eel sheds.

The oldest eel sheds are from the eighteenth century. There were at least a hundred along this thirty-mile stretch of coast once, and fifty or so are still standing. They are typically named after the fishermen who used them or the myths and legends said to have taken place in them. They're called things like the Brothers' Shed, Jeppa's Shed, Nils' Shed, the Hansa Shed, the Twin Shed, the King's Shed, the Smuggler's Shed, the Tail Shed, the Cuckoo Shed, and the Perjurer's Shed. Some of the sheds are derelict, some have been converted into seaside summer cottages, but a handful are still used for their original purpose. It's in these sheds you find a second category of people, quite distinct from the natural scientists, who have historically had a close relationship with the eel: eel fishermen.

Here, on the Swedish eel coast, only a few remain, and it's a shrinking brotherhood, but their presence and profession have shaped life in this part of the world for a very long time. For centuries, eel fishing has been central to the area's culture, traditions, and language. Here, almost everyone knows the old eel fishermen by name. Here, most have at one time or another attended an eel feast, the special late-summer or early-autumn celebrations dedicated to the eel. Here, the eel, the traditions built around it, and the knowledge about it, have become an intrinsic part of the local identity.

And it has been thus since at least the Middle Ages. Fish-

ing along the eel coast is organized through the distribution
of a special kind of fishing rights, called *åldrätter.* The word
drätt comes from the Swedish verb for "pull" and refers to the
fishing technique normally used here. It's an ancient system,
one with roots in a feudal, predemocratic time, and the only
place where it survives is here, on the Swedish eel coast. The
system stems from a time when Skåne was still part of Denmark; the oldest extant documentation about it dates from
1511 and tells us that a certain Jens Holgersen Ulfstand of
Glimmingehus purchased two *åldrätter* from the archbishop.
The rights were sought after, because eels were a plentiful
and popular food. When Skåne became Swedish in 1658, the
Swedish king appropriated the local fishing rights and redistributed them in accordance with his authoritarian Swedifica-
tion policy to members of the clergy and nobility in exchange
for loyalty. The owners of *åldrätter* could, in turn, make lu-
crative deals leasing those rights to fishermen and farmers.
And thus, the eel has also been a tool for exercising power.

The eel feast is a leftover from those days. The Swedish
word for it, *gille,* comes from the word *gäld,* meaning "debt"
or "payment," and refers to the fee a fisherman would have to
pay for his fishing rights. The payment would usually be due
at the end of the eel season and was made in actual eels. And
thus, the eel also served as a kind of currency.

A traditional eel feast typically requires at least four different eel dishes; there are many local specialties. Fried eel,
boiled eel, and eel soup. Smoked eel cleaned and soaked in
brine overnight before being scalded and smoked over alder
wood. So-called *luad* eel, which is lightly salted, put on a spit,
and then baked in a hot oven, making it smoked and roasted

at the same time. *Halmad* eel, which is a large eel cut into portion-sized pieces and fried in a hot oven in a pan filled with rye straw. *Pinna* eel, smaller eels salted and fried with alder sticks and juniper brush. Sailor's eel, which is smoked eel braised in dark beer and fried in butter. *Fläk* eel, cleaned, deboned, oven-baked eel stuffed with dill and salt. And in this way, the eel has become the focus of a unique food culture.

The eel coast is divided into a total of 140 *åldrätter*. They range from five hundred to one thousand feet in width and extend a few hundred feet into the sea. Only the owner or leaser of an *åldrätt* can fish for eels in that particular location. The eel sheds were built adjacent to the designated *åldrätt* areas. They were small, simple houses, with a storage room and a small living space with a table and a few cots for sleeping. During the fishing season, the fishermen typically lived in them in order to guard the corves where the caught eels were kept, or to ensure they would be ready to head out and salvage their equipment in case of a storm. Before the sheds were built, fishermen would simply turn their wooden boats over on the beach and sleep under them on simple beds of straw.

The season traditionally lasts only three months, the length of the so-called eel darkness, when the eels move out into the ocean, passing along the coast on their way to the Sargasso Sea. These eels—the largest, fattest ones, which have adapted their bodies to the long journey across the Atlantic—are the ones the fishermen are after. Usually at the end of July, the fishermen place the traps they will then inspect every day at dawn until the start of November, when they are removed. That's the end of the season. No more eel darkness.

Eel fishing has always been a cottage industry. Neither the

location nor the eel itself has permitted scaling up. The fishing is primarily done using a so-called *homma*, a special kind of trap equipped with a grapnel and floaters, which has long mesh wings leading into a tapering bag in which the caught eels are collected. The boats used are small and flat-bottomed to aid navigation in shallow water and to facilitate their being pulled up onto the beach. Both *hommor* and boats are traditionally crafted by the fishermen themselves.

Things do change, of course, but only in minor ways. The boats, which used to be made of tarred oak, are now plastic. Where oars were once used, people now prefer outboard motors. Fishing rights are no longer paid for in eels and no longer passed down from father to son. These days, women are allowed in both eel sheds and at eel feasts. But other than that, things are done the way they've always been done. Partly because the eel demands it and partly because that's how the fishermen want it, but also because on the eel coast people agree that there's value to keeping traditions and knowledge alive. And thus, the eel has, in time, become a cultural heritage.

WHAT KIND OF PERSON CHOOSES TO BECOME AN EEL FISHERMAN? What does the eel provide such a person with? A profession and an income is the simple answer. But that's not the whole story. True, the eel has been an important source of food in large parts of Europe throughout history, but it has always been tricky. Difficult to catch, difficult to understand, enigmatic, and to many people simply unpleasant. It has forced fishermen to develop special methods and tools; its peculiar behavior has kept the fishing industry small scale even though demand has been high. It can't be farmed like salmon, for instance; in fact, it won't breed in captivity at all. As a source of nourishment, the eel has been crucial for a lot of people, but it has rarely been particularly cooperative. And today, when fewer and fewer people eat eel and catches are shrinking, why become an eel fisherman at all?

If you were to ask the people on Sweden's eel coast, many would probably tell you it's rarely a choice. You're born into it; you have been groomed for it over the course of generations. It goes without saying there are no university courses or professional training programs for eel fishermen. The special knowledge an eel fisherman possesses isn't gained in the classroom or a laboratory. It has been passed down through centuries, like an ancient story that no one has ever bothered to write down. How to craft a *homma* or how to flay an eel, how to read the sea and the weather and how to interpret the eel's movements under the surface: this specific and particular knowledge has been transmitted through practical work, as a shared experience transcending the ages. And thus, fishing for eel has often been a profession that runs in families, handed down from one generation to the next. No one becomes an eel fisherman who doesn't have it in his or her blood. And no one becomes an eel fisherman who doesn't also view the work as a way of protecting and preserving something bigger than fishing per se: a cultural heritage, a tradition, and knowledge.

The parts of Europe where eel fishing has been most important have rarely included big, well-known cities. The metropolises of the eel are not those of humankind. Instead, they have been peculiar places, populated by peculiar people. Stubborn and proud people who, like those on the Swedish eel coast, have often inherited their profession from their fathers and been shaped by hard labor and simple circumstances. Who have let their work become their identity and who have as a consequence, much like Johannes Schmidt, kept plying the waters in their boat, hunting for eels even when common sense told them not to. Oftentimes, these people have nurtured a kind of outsider status and a suspicious attitude toward the powers that be. The eel fisherman has, in more places than the Swedish eel coast, been a creature apart.

GLASS EELS ARE FISHED IN THE RIVER ORIA IN THE SPANISH BASQUE Country in winter and early spring. The river, which empties into the Bay of Biscay, meanders through the mountainous Basque landscape and is a popular thoroughfare for the transparent glass eels, which after a couple of years of drifting across the Atlantic, swim up waterways to find a home for the next ten, twenty, or thirty years. Many of them don't make it very far. Near the estuary by the coast, fisherman spend cold, rainy nights in wooden boats, sieving the fragile eels out of the water.

The small village of Aguinaga, located on the river a few miles inland, has only six hundred residents but no fewer than five companies that catch and sell glass eels. Here, too, professional knowledge is ancient and inherited. The glass eels come in on the tide on cold nights, under a full or crescent moon and preferably when the sky is slightly overcast. They float near the surface in massive shoals, like enormous, silvery tangles of seaweed; the fishermen glide slowly back and forth in their boats; the light from the lanterns at their prows is reflected in the living blanket of fish. They lift the glass eels out by hand, with round nets attached to long rods.

The glass eel is considered a delicacy in the Basque Country, and only there these days. The tradition of consuming the eel in this frail, transparent state has, however, historically been widespread. In the United Kingdom, glass eels were once caught in the Severn. They were fried whole while still alive together with a bit of bacon, or with a beaten egg in a kind of omelet, a so-called elver cake. In Italy, glass eels used to be caught in the Arno River in the west and around Comacchio in the east. There the preferred way of serving them was boiled in tomato sauce with a sprinkle of parmesan. Eating glass eels was also popular in some parts of France. These days, however, it's a dying tradition. As the number of glass eels wandering up Europe's rivers has plummeted, the fishing industry built around them has also ceased to exist. It's really only the Basques who stubbornly refuse to give up.

There are, of course, rational reasons for this. First on the list are financial concerns. Glass eels have been fished here for a long time. It's said they used to drift up the Oria in such great quantities that farmers would catch them from the banks by the netful and feed them to their pigs. But it's their scarcity, the increased threat to their existence, that has ultimately made the glass eel a more sought-after and exclusive delicacy, in a twist of logic that is unique to humankind. In the Basque Country it's eaten fried in the finest olive oil with a hint of garlic and mild chili. It's served burning hot in a small ceramic dish; diners eat it with a special wooden fork to avoid burning their lips. In peak season, a small portion, 250 grams, can cost sixty or seventy dollars in the finer eateries in San Sebastián.

But the eel fishermen in Aguinaga and along the Oria have other reasons to continue with their trade. They simply don't want to stop. Because they feel it's their right. Because this was precisely what their ancestors did before them and because this particular way of fishing for eel is, aside from a way to earn a living, what makes them who they are. The region is also a stronghold of the Basque separatist group Euskadi Ta Askatasuna. People here are used to being self-reliant. For forty years they were marginalized and oppressed under the Spanish dictator Francisco Franco, so they remain vigilant against power grabs by bureaucrats in Madrid or Brussels. Here, the fishermen will return to the river with their nets and lanterns no matter what politicians and scientific experts have to say about it. Until the very last eel fisherman is gone. Or the very last eel.

AROUND LOUGH NEAGH IN NORTHERN IRELAND, LOCALS HAVE fished for eels for at least two thousand years; the eels caught there are often described as Europe's finest. Lough Neagh is found at the northeast corner of Ireland. It is the largest lake in the British Isles, located west of the Mourne Mountains in a fairly barren landscape; for large parts of the year, it is characterized by a rather unforgiving climate, prone to severe storms. Yet even so, the fishing here continues much as it always has. Because that is what each successive generation has been taught to do. Because neither the location nor the eel has allowed any variation.

In Lough Neagh, the catch is primarily yellow eel, and the tool used is a spiller. Long lines with multiple hooks baited with worms are set from simple boats. Two fishermen per boat will set four spillers with four hundred hooks on each every day during peak season. Sixteen hundred hooks that need to be baited by hand and checked at the crack of dawn when the cold and fog turn fingers into stiff glass rods.

Traditionally, the catch has been shipped to London. Eel was for a long time a popular food in the capital, sold in little shops and market stands. It was eaten fried with mash, or as jellied eel, sliced rounds of eel boiled in a stock that sets into jelly. It was considered good-value-for-the-money everyday fare, and was intimately associated with the working class of the East End. The eel was fatty and rich in proteins and significantly cheaper than meat, which is why it was sought after by the poor and predictably often despised by the wealthy.

But Londoners' fondness for eel was not the only reason Lough Neagh eels ended up in London. There were political reasons as well. When the British colonized large parts of Ireland in the sixteenth and seventeenth centuries, they confiscated not only the most fertile land but also valuable natural resources. In 1605, the Irish locals around Lough Neagh were forced to give up their fishing rights, and for more than three hundred and fifty years, the fishing was controlled by the English colonizers. Wealthy Protestants decided how many eels were to be caught, what was to be done with them, and how much fishermen would be paid for them. The fishermen, often Catholic farmers forced from their land, obliged to find other ways of making a living, were poor and powerless. The eel was an emergency solution to stay alive.

For several hundred years, all fishing rights were in the possession of the Earl of Shaftesbury, but in the mid-twentieth century, they were sold to a consortium called the Ring, which consisted of a handful of wealthy eel merchants in London. The Ring still controlled all eel fishing in Lough Neagh when a group of Catholic fishermen banded together in 1965, forming the Lough Neagh Fishermen's Cooperative Society. Together, the cooperative was able to raise the money to buy 20 percent of the lake's fishing rights. In the years that followed, more money was set aside and the remaining 80 percent was purchased as well. That this happened at the same time the Troubles broke out was, of course, no coincidence. The members of the Ring testified to being forced to sell their shares under threat of violence; they also testified that the consortium's ships had been attacked. It was said the eel fishermen were, to a man, members of the Irish Republican Army.

And thus, the eel became embroiled in the violent Northern Irish conflict, which has always had as much to do with class, power, ownership, wealth, and poverty as it has with religion. Today, fishing on Lough Neagh is 100 percent controlled by the Lough Neagh Fishermen's Cooperative Society, and those who still fish for eel are not about to forget where they came from. Stubborn pride drives them to keep baiting their hooks and setting their spillers. Because that's what's always been done and how it should be.

AND NOW ALL THIS WILL DISAPPEAR. THE CULTURAL HERITAGE AND the traditions. The regional dishes and landmarks. The eel sheds, boats, and fishing tools. The knowledge that has been

passed down the generations. And eventually, the memory itself of all those things.

At least that's the fear, on the shores of Lough Neagh and
in Basque Aguinaga, and on the Swedish eel coast. Because
as the eel population shrinks, the calls to protect grow stronger. Fishing for glass eels is already banned in many parts of
the continent. Scientists and politicians are working toward a
complete ban across Europe.

So be it, the fishermen say, but remember that you're not
just robbing us of our livelihoods. Traditions, knowledge, and
a valuable, old cultural heritage will also inevitably be lost.
More than that, they claim, humanity's relationship with the
eel is at stake. If people can no longer fish for eel—catch it, kill
it, and eat it—they will lose interest in it. And if people have
no interest in the eel, it's lost anyway.

That's why the Lough Neagh Fishermen's Cooperative
Society is now working as hard to save the eel as to catch it.
Among other things, it runs an extensive and costly project
to buy and release glass eels into the lake. The eel fishermen
on the Swedish eel coast have organized and are working to
increase awareness of the plight of the eel as well. They have
founded something called the Eel Foundation, which, much
like the fishermen at the Lough Neagh society, works to release
eels in order to bolster stocks. In 2012, the Cultural Heritage
Association of the Eel Coast was established, with the aim of
getting eel fishing and its traditions in Sweden declared an
intangible cultural heritage. On its website, the association
writes: "A total ban on eel fishing means a living culture, a
local craft, and a unique culinary heritage becomes history.
The eel sheds along the coast will be turned into summer

homes for the wealthy. The stories will fall silent. The interest in the eel, and thus the eel itself, will be lost."

This is the great paradox, which has also become part of the eel question of our time: in order to understand the eel, we have to have an interest in it, and to have an interest in it, we have to continue to hunt, kill, and eat it (at least according to some of the people who, after all, are closer than most to the eel). An eel is never allowed to simply be an eel. It's never allowed to just be. Thus, it has also become a symbol of our complex relationship with all the other forms of life on this planet.

10

Outwitting the Eel

One summer, we tried to *klumma*. It's an old fishing method used in streams in rural Skåne, in southern Sweden. By all accounts, it's an activity that belongs to a different world, since the method itself is so insane it's hard to imagine how anyone would be capable of inventing it today. But somewhere, at some point, someone did, and also discovered, against all odds, that it not only worked but was highly effective. Somehow, this knowledge then spread, in patterns that are both undiscernible and inexplicable, to finally arrive at my dad, who in turn passed it on to me, as though it were the most natural thing in the world.

Which it is not. When you *klumma* for eel, you thread a needle with a long piece of extra-strong sewing thread and hold it in one hand while you hold a worm in the other. You stick the needle through the worm, pull the thread all the way through and repeat until you have several feet of worms,

which you roll into a quivering, stinking ball of slime and se-
cretions and writhing bodies. You then attach a sinker and a
line to the ball, but no hook.

You fish at night, preferably from a boat. The ball of
worms is thrown into the water and left to settle on the bottom, while you hold the taut line gently. When the eel finds
the ball and bites into it, you respond with an immediate tug.
If you're skilled enough, and since the eel's tiny but slightly
curved teeth make it cling to the thread in a slightly hangdog
way, you can pull the eel into your boat in one quick, smooth
motion. At least in theory.

Dad had never tried it before. He hadn't even seen anyone
do it. But we both realized it would take, first and foremost,
a very large number of worms. Dad had an idea about how
to find them. He told me to water the lawn while he grabbed
a pitchfork, cut off a piece of electrical cord, attached one of
the exposed wires to the prongs, and shoved the fork into the
ground.

"You'd better stand back now," he said. "And put your
wellies on."

I stood on the front steps in my boots, pulse racing, watching as he plugged the cord in and two hundred and twenty
volts jolted through it, into the pitchfork, and down into
the damp soil. At first, nothing happened, not a sound, not
a movement. Then the worms started appearing out of the
ground, hundreds of them, covered in dirt, wriggling in dis-
tress. The whole lawn looked like one big living organism.

Once dad had turned the power off, we walked around,
picking up our bait. It took just ten minutes to fill a big jar.

WHEN NIGHT FELL, WE WERE IN OUR WOODEN BOAT, HOLDING THE line with the revolting ball of worms dangling in the water beneath us, and I wondered what the point was. What was the *point* of this fishing method? Of course, one person may find meaning where another can't even discern sense, but doesn't meaning have to be part of a context? And doesn't this context have to at least be perceived as bigger than oneself? After all, people have a need to be part of something lasting, to feel that they are part of a line that started before them and will continue after they're gone. They need to be part of something bigger.

Knowledge can, of course, be the bigger context. All kinds of knowledge, about crafts or work or ancient insane fishing methods. Knowledge can, in and of itself, constitute a context, and once you become a link in the chain of transmission, from one person to another, from one time to another, knowledge becomes meaningful in itself, quite apart from considerations of utility or profit. It's at the heart of everything. When you talk about human experience, you're not talking about individual experience, you're talking about our communal experience, which is passed on, retold, and reexperienced.

But this particular knowledge—how to string worms up on thread in order to try to trick an eel—was there any meaning to that anymore? And this particular experience—sitting silently in a boat at night, with a ball of slowly dying worms on a line beneath you—was there any humanity left in that?

Before long, it was completely dark and we were sitting dead still. The only sound was the gentle rushing of the water around us; from time to time, we'd raise our hands, lifting the ball of worms up off the bottom with a soft tug. As if to let whatever was moving down there know we were there.

And it soon returned the favor. A short, distinctive yank that felt like a sudden slap in my hand.

I instinctively raised my hand straight up in the air and saw the ball of worms rising toward the surface and in its wake, a large eel, slithering eagerly this way and that as though frantically swimming toward me instead of trying to escape. I pulled it out of the water and over the railing and then there it was, lying by our feet, whipping its head from side to side, a sudden reminder of the consequences of my action.

It was over in seconds, and then it started again. We caught twelve eels that night. Another night a few days later, we caught fifteen. They kept biting and we kept pulling them into the boat, like pulling carrots out of a vegetable patch. It was as though there were an endless source of eels that had suddenly opened just for us; it was, if not meaningful, at least comprehensible; the method, the knowledge, was functional and apparently even effective. We had found a way to outwit the eels that was in a different league from any other method we'd ever tried.

And yet, we never *klummade* again after those two nights. I think it had to do with the images it evoked. The yellowish-brown, shiny eel, slithering through the sediment in the dark, biting into a quivering mass of dying worms, letting itself get hauled out of the water, with neither hook nor struggle, as though it had given up; as though it were trying to escape something in the depths. It didn't tally with what we wanted the eel to be. The eel didn't behave as we expected it to. Maybe we had gotten too close to it.

11

The Uncanny Eel

On November 11, 1620, the *Mayflower* dropped anchor off Cape Cod in the southeast part of present-day Massachusetts. Just more than two months earlier, the ship had left England with 102 passengers and about thirty crew. The passengers were mostly Puritans, members of a strict Protestant church that preached an orthodox, ascetic form of Christianity. They had left England as a result of both poverty and religious persecution, first for temporary exile in the Netherlands, then to the west to start over in the New World. They left not only because they hoped to find freedom and prosperity in this new land, but also because they believed it was God's will. Rather than refugees, they thought of themselves as chosen. Chosen by God to be saved, chosen to spread the one true doctrine across the world in His name.

Salvation, as it so often happens in Christian stories, would, naturally, come only after a series of trials. And when it finally came, it did so in an unexpected form.

It was already winter when the *Mayflower* reached the coast of North America. The land was cold and desolate; most of the passengers were forced to remain on the ship for months before they could leave. The smaller expedition that rowed ashore on the first day to do reconnaissance had a bad time of it. Several of them froze to death as they camped overnight on the snowy beach. The survivors were cheered to discover a cemetery and some seemingly abandoned winter stores of corn and beans, but after sacking the stores, they found themselves hunted by the natives whose food they'd stolen. One night, they were attacked by warriors with bows and arrows and narrowly escaped.

Tuberculosis, pneumonia, and scurvy soon broke out aboard the ship. Food was scarce and the water dirty. When spring finally arrived, only 53 of the 102 passengers were still alive. Half the crew had perished as well.

It was March before the surviving colonizers were able to leave the ship at last, still determined to follow through on their plan and fulfill the will of God. They were famished and frozen and had in the way of possessions not much, other than their conviction that God was on their side. They didn't know where they should start building their colony or how they could make peace with the natives. Nor did they know where to hunt, which plants were edible, or how to find potable water. The promised land could perhaps be welcoming, but clearly only to those who understood it.

That's when they came across Tisquantum. A member of the Patuxet tribe, he had been captured by the English years earlier, taken to Spain, and sold as a slave, before managing to escape to England, where he learned the language.

Eventually, he boarded a ship back to North America, only to find that his entire tribe had been wiped out by an epidemic probably brought by the English.

There was no apparent logic to his actions, and a person's motives cannot always be explained by his backstory, but by all appearances, Tisquantum saved the imperiled English colonizers. One of the first things he did was gift them an armful of eels. After their very first meeting, Tisquantum went down to the river, and "at night, he came home with as many eels as he could well lift in one hand, which our people were glad of," noted one of the pilgrims in a diary later sent back to England. "They were fat & sweet, he trod them out with his feete, and so caught them with his hands without any other Instrument." It was a gift from God in their hour of need, the salvation they had never stopped praying for.

Before long, Tisquantum had taught the pilgrims how to catch eels and where to find them. He also gave them corn and taught them how to cultivate it; he showed them where they could find wild vegetables and fruits and advised them on how and where to hunt. Not least, he helped them communicate with the local natives and was key to negotiating the peace agreement that was pivotal to the lost Englishmen's future in America.

And thus, the pilgrims survived, becoming, in time, legends in the American creation myth. The *Mayflower*'s arrival has been a symbolic and epoch-making event in American history ever since, mythologized and romanticized in countless patriotic contexts.

In November 1621, a year after their arrival and around the date that has ever since, and because of the pilgrims' survival, been called Thanksgiving, they wrote in their diaries about the amazing land they had found. They wrote about the grace that had, after all their tribulations, been extended to them and thanked the Lord for all the trees and plants and fruits surrounding them, for the animals and fish and fertile soil and, of course, for the eels they "effortlessly" fished out of the river in great quantities every night.

It would have made complete sense for the eel to have become an important figure in American mythology, a fat, shiny symbol of the promised land, the gift that sealed what was preordained. But that didn't happen. Perhaps because the eel's nature doesn't lend itself well to solemn symbolism. Perhaps because it soon became associated with the simple eating habits of the poor rather than with feast days. Perhaps also because the gift had come from a native man.

For some reason, this gift from God to the early pilgrims has been all but erased from the grand narrative. The story of the colonization of North America is full of myths and legends, but the story of the eel isn't one of them. On Thanksgiving, Americans eat turkey, not eel, and other animals—buffalo, eagles, horses—have been the ones to shoulder the symbolic weight of the patriotic narrative of the United States of America. True, the colonizers continued to catch and eat eels, and by the end of the nineteenth century the eel was still an important ingredient in the American kitchen. But it gradually disappeared from dinner tables. After the Second World War, the eel's reputation lay in tatters, and by the end of the 1990s, eel fishing had more or less completely ceased along the East Coast. Today, many Americans think of the eel as a troublesome, fairly unappetizing fish they want as little to do with as possible. Sometimes, even the gifts of God are only begrudgingly accepted.

THIS UNCERTAIN, CONTRADICTORY ATTITUDE TOWARD THE EEL WAS, of course, not unique to the arrival of the *Mayflower* in North America. Throughout history, the eel has aroused ambiguous feelings in the people who have encountered it. At times reverence, but also an inevitable unease. Curiosity, but also rejection.

In ancient Egypt, the eel was considered a mighty demon, an equal of the gods and a forbidden food. A creature moving effortlessly beneath the glittering surface of the holy Nile, slithering through the sediments of existence itself. Archaeologists have found mummified eels in tiny sarcophagi, laid to their eternal rest next to bronze statuettes of the gods.

Granted, many animals symbolized divinity in ancient Egypt. The sun god Ra was often depicted with the head of a falcon. The god of the Underworld, Anubis, had the head of a jackal. Thoth, the god of wisdom, was given the head of an ibis. The goddess of love, Bastet, had a woman's body and a cat's head. Every animal represented different characteristics, of course, but the blurring of the line between human and animal was also in itself a sign of divinity. Atum, who in Heliopolis was the father of all other gods and pharaohs, was also the god associated with the eel. In one depiction, Atum has a human head, a pointy beard, and a crown signifying his divine status, and behind a wide, intimidating cobra shield, his body is that of a long, slender eel, complete with realistic fins. The human head and eel's body together symbolized a kind of wholeness, the union of positive and negative forces.

In ancient Rome, opinion was also divided when it came to the eel. Some refused, like the Egyptians, to eat eel, not because it was holy but rather because it was considered unclean and loathsome. Perhaps because eels were often caught near sewer outlets. Perhaps because dried eel skins were used to make a kind of belt to discipline disobedient children.

Many Romans seem to have preferred the conger (*Conger conger*) or the moray eel, which is related to the eel—but whatever the species, the eel was often associated with something dark and macabre. Both Pliny the Elder and Seneca the Younger describe how the Roman military commander Vedius Pollio, a friend of Emperor Augustus, had the habit of punishing slaves by throwing them into a pool filled with eels. The bloodthirsty fish ate their fill and were then served to Vedius Pollio's guests as a particularly fatty and luxurious delicacy.

A FISH, BUT ALSO SOMETHING ELSE. A FISH THAT LOOKS LIKE A snake, or a worm, or a slithering sea monster. The eel has always been special. Not least in Christian tradition, in which the fish has been, from the beginning, one of the most central symbols, the eel has been viewed as a thing apart.

It's said the earliest Christians, during the first century after the birth of Christ, used the fish as a secret sign. Since Christians were persecuted in many places, a level of caution was required, so when two believers met, one would draw an arced line on the ground. If the other drew a similar one from the other direction, the lines together formed a stylized fish, and the two knew they could trust each other. This symbol can be found in the catacombs of Saint Calixtus and Saint Priscilla in Rome, dating back to the very first centuries of the Common Era.

The fish was significant for several reasons. Long before the birth of Christianity, it had been a symbol of luck in Mediterranean culture. With the coming of Jesus, the fish also became a symbol of revivalism and confession. "Follow me, and I will make you become fishers of men," Jesus says to the very first apostles, Simon and Andrew, in the Gospel. Newly saved people are called "small fry," and in the Gospel, Jesus likens entering the kingdom of heaven to fishing: "The kingdom of heaven is like a net that was let down into the lake and caught all kinds of fish. When it was full, the fishermen pulled it up on the shore. Then they sat down and collected the good fish in the baskets but threw the bad away. This is how it will be at the end of the age. The angels will come and separate the wicked from the righteous."

The fish also plays a well-known role in the stories of the miracles of Jesus, including the miracle of the loaves and fishes, when he feeds five thousand people with only two fish and five loaves of bread. Or when the resurrected Jesus reveals himself to his apostles by Lake Tiberias and provides them fish to eat, convincing them that it's really him. The Greek word for fish, *ichthys*, has also long been read as the acronym *Iesos Christos Theou Yios Soter*—"Jesus Christ, Son of God, the Redeemer."

But that's all about fish, not eels, and many things point to early Christians making a distinction between the two. All the good things the fish came to represent in the Christian tradition were reserved for species other than the eel. The eel was no fish; it was something else. And even if the eel had been considered a fish, it was not a fish like the others. It didn't possess the usual characteristics of a fish. It didn't look or behave like fish normally do.

This is clear if you read between the lines in Leviticus, in which God's opinions about all aquatic creatures are clearly expressed:

> *These you may eat, of all that are in the waters.*
> *Everything in the waters that has fins and scales,*
> *whether in the seas or in the rivers, you may eat. But*
> *anything in the seas or the rivers that has not fins*
> *and scales, of the swarming creatures in the waters*
> *and of the living creatures that are in the waters, is*
> *detestable to you. You shall regard them as detestable;*
> *you shall not eat any of their flesh, and you shall detest*
> *their carcasses. Everything in the waters that has not*
> *fins and scales is detestable to you.*

What God apparently means to say, assuming the word choices and repetitions are correctly interpreted, is that fish and other aquatic animals without fins and scales are abhorrent. They mustn't be eaten; they're uncanny; they shall be viewed with loathing. And at least in the Jewish reading of God's intentions, that means the eel is detestable. It's not considered kosher, and its smooth, slimy body consequently has no place on the Jewish dinner table.

Now, this is all a misunderstanding, of course, sort of like when Leviticus also lumps bats in with birds. The eel has both fins and scales. They're just a bit difficult to make out, especially the scales, which are so incredibly small and covered in slippery slime that they're almost imperceptible to the touch. But it is a misunderstanding that shows that when it comes to eels, not only are science and the eel itself suspect, you can't trust God either. Or God's interpreters. Or words.

BE THAT AS IT MAY, THE EEL REMAINED DETESTABLE, IF NOT TO ALL then to many, and if not as a food or a cultural heritage, then at least as a metaphor. Even beyond fallacies and religious misunderstandings, it has, at times, come to represent the unwelcome. Whatever is strange and unpleasant to us. What may have to be allowed to exist, out of view, but which should not be allowed to reach the surface too often.

In one of the twentieth century's most memorable scenes from literature, a man is standing on a beach, pulling on a long rope that stretches out to sea. The rope is covered in thick seaweed. He yanks and tugs, and out of the foaming waves comes a horse's head. It's black and shiny and lies there at the water's edge, its dead eyes staring while greenish eels slither from every orifice. The eels crawl out, shiny and entrails-like, more than two dozen of them; when the man has shoved them all into a potato sack, he pries open the horse's grinning mouth, sticks his hands into its throat, and pulls out two more eels, as thick as his own arms.

This macabre fishing method is described in Günter Grass's 1959 novel, *The Tin Drum*. Rarely has the eel been more detestable.

The eel does not appear frequently in literature or art, but when it does, it's often an unsettling, slightly revolting creature. It's slimy and slithering, oily and slippery, a scavenger of the dark that salaciously crawls out of cadavers with gaping mouth and beady black eyes.

Sometimes, however, it's more than that. In *The Tin Drum*, the eel actually plays a rather important role. It both foreshadows and triggers tragedy.

The people standing on that Baltic beach, watching the man pull the horse's head from the sea, are the novel's main characters, the boy Oskar Matzerath; his father, Alfred; his mother, Agnes; and her cousin and lover, Jan Bronski. Agnes is pregnant but hasn't told anyone. We don't know who the father is, Alfred or Jan, nor do we know if Alfred is really Oskar's father. Agnes is depressed and self-destructive and seems to view the life growing within her more as a devouring tumor than a gift. What's happening inside her is a mystery, to both her family and the reader.

The four of them have gone for a walk along the beach when they come across the eel fisherman. Agnes curiously asks what he's doing, but he makes no reply. He just grins, flashing filthy teeth, and continues to tug on the rope. Once the horse's head is out of the water and Agnes sees the eels crawling out of its skull, something happens to her. She's revolted by them both physically and psychologically. She has to lean against her lover, Jan, to keep from swooning. The seagulls swarm above them, flying in ever-tighter circles, screeching like sirens; when the grinning man pulls the two fattest eels out of the horse's throat, Agnes turns and vomits. It's as though she's trying to expel both her acute nausea and the unwanted fetus in her belly. As though one is inextricably linked to the other. She never fully recovers from the experience.

Jan eventually leads Agnes away down the beach; Oskar and Alfred stay behind, watching the man pull the last enormous eel, sticky with white, porridge-like brain substance, out of the horse's ear. Eels don't just eat horses' heads, they eat human bodies, too, the man explains, and tells them the eels grew especially plump after the Battle of Skagerak during the First World War. Oskar stares, mesmerized, his white tin drum slung around his neck and resting on his belly. Alfred is thrilled and promptly buys four eels from the man, two large and two medium ones.

The event on the beach changes Agnes. The sight of the slithering eels and the grotesque horse's head awakens something in her. She grows increasingly ill and tries to manage her condition with food. She eats constantly, binging and vomiting by turns. What she eats is fish, and eel in particular. She devours fatty pieces of eel swimming in cream sauce, and when her husband refuses to serve her more fish, she goes to the fishmonger and returns with an armful of smoked eels. She scrapes the skin clean of fat with a knife and licks it, then vomits once more. When her husband, Alfred, nervously asks if she is pregnant, she only snorts with derision and serves herself another piece of eel.

Agnes dies shortly after. Its unclear if she eats herself to death, or if perhaps her heart has broken. At her funeral, her son, Oskar, studies her in the open casket. Her face is haggard and slightly jaundiced. He pictures her suddenly sitting up and vomiting once more, imagines there's still something inside her that has to come out, not just an unwanted child but also that alien and detestable thing that in such a short time devoured and killed her. Which is to say, the eel.

"From eel to eel," Oskar thinks, standing by the coffin, "for eel thou art, to eel returnest."

And when his dead mother doesn't sit up and vomit, he experiences relief and closure. "She kept it down and it was evidently her intention to take it with her into the ground, that at last there might be peace."

It's a devastating metaphor. The eel as death incarnate. Or rather, not just death but also death's opposite. The eel as a kind of symbolic link between beginning and end, between the origin of life and its demise. Ashes to ashes, eel to eel.

IN THE MID-TWENTIETH CENTURY, WHEN *THE TIN DRUM* WAS FIRST published, science had teased out most of the eel's secrets. It had been demystified and rendered comprehensible. Humanity had slowly but surely homed in on the answer to the eel question. Its origin had been found and its method of reproduction established. Progress had been slow, like a snail next to the bullet train of scientific advancement that had taken place since the Renaissance, but the eel was now for the most part understood. No longer limited to simply pointing to its undeniable existence, we were in a position to discuss the features of that existence. We knew not only that the eel *is*, we also knew some of *what* the eel is. We no longer had to rely solely on faith.

And yet the eel continued to be associated with the irrational psyche of humankind, with the alien and unfathomable, in both literature and art. It remained a slimy, frightening creature of the dark, slithering out of the depths. A creature unlike others.

In Fritiof Nilsson Piraten's Swedish classic *Bombi Bitt and Me*, from 1932, the eel is a devil, a horned monster that has grown to more than ten feet long over the course of countless years in the depths. In a remote and possibly bottomless Scanian pond, it has hidden away from humanity, until the book's main characters, Eli and Bombi Bitt, along with old man Vricklund, set out to catch it one night. Vricklund manages to pull it out of the pond; it's a "dark, monstrous creature, that whipped the water to foam"—and then a wild wrestling match ensues. The eel rises up like a "living telephone pole"; the moonlight outlines its large horns; the struggle ends only when Vricklund sinks his teeth into its enormous body.

"I bit that bastard to death," Vricklund declares, blood still dripping from his mouth. But it's a temporary victory. The eel is resurrected. It comes back to life with a heavy sigh, slithers away through the grass, and disappears into the underworld through a hole in the ground. Back to the place it evidently came from, the shadows, the subconscious, the lowest, darkest circles of the soul.

In Boris Vian's surrealist love story *The Foam of Days*, from 1947, the eel is an absurd creature that foreshadows impending tragedy. It emerges from the kitchen faucet at the very start of the story. Every day, it pokes its head out of the tap, looks around, and vanishes again. Until, that is, a crafty cook one day places a pineapple on the kitchen counter, and the eel, unable to resist, sinks its teeth into it. The cook makes a delicious eel pâté, which the protagonist, Colin, eats, thinking of his love, Chloé, whom he has just met and is set to marry, but who will soon fall gravely ill. A water lily is growing inside her chest, an aquatic plant from the world of the eel. It grows like an aggressive tumor, killing her and leaving Colin heartbroken and alone.

The eel's greatest performance, at least in literature, however, is in the 1983 novel *Waterland* by the English author Graham Swift. It tells the story of Tom Crick, a history teacher who tries to capture the imaginations of his bored and scientifically minded students with stories about himself and his childhood. He examines his own unreliable memory, trying to understand why things ended up the way they did. His marriage to Mary and their childlessness. Her nascent insanity. Where did it all start? Maybe with the live eel another boy stuck down her pants when she was a child?

Or with his brother Dick, who also wooed Mary when they were young and who won a swimming competition just to impress her? Like an eel on its way to the Sargasso Sea, he swam farther than anyone else in order to reach his goal—the goal that is also the goal of existence. But why did he? And what does it really mean?

The story is vague and unreliable. Who really knows what the truth is? But the eel is ever present. From start to end. It slithers through the entire story like a constant reminder of everything that is hidden or forgotten.

And toward the end, Tom Crick tells his students about the eel itself. About the eel question and its scientific history, with all its guesswork and mysteries and misunderstandings. About Aristotle and the theory of the eel springing from mud. About Linnaeus, who thought the eel was self-propagating. About the famous Comacchio eel, about Mondini's discovery and Spallanzani's questioning of it. About Johannes Schmidt and his dogged search for the eel's birthplace. About the curiosity that drove them all. This is what the eel can teach us, Tom Crick argues. It tells us something about the curiosity of humankind, about our unquenchable need to seek the truth and understand where everything comes from and what it means. But also about our need for mystery. "Now there is much the eel can tell us about curiosity—rather more indeed than curiosity can inform us of the eel."

BUT WHY IS THE EEL CONSIDERED SO UNPLEASANT? WHY DOES IT arouse those kinds of feelings in us? Surely it's not simply because it's slippery and slimy, or because of what it eats, or because it likes the dark? Nor can it be based solely on religious misinterpretations. No, it's probably also because it's secretive, because there seems to be something hidden behind its apparently lifeless black eyes. On the one hand, we've seen it, touched it, tasted it. On the other hand, it's keeping something from us. Even when we get really close to it, it somehow remains a stranger.

In psychology, and in art, there's a particular kind of unpleasantness referred to as uncanniness. The German psychiatrist Ernst Jentsch wrote an article in 1906 entitled "Zur Psychologie des Unheimlichen," in which he defines the concept of the *unheimlich*, the uncanny, as "the dark sense of insecurity" we are overcome with when we encounter something new and strange. What frightens us, Jentsch explains, what's uncanny, is that which makes us intellectually unsure, what lack of experience or the limitations of our senses prevents us from immediately recognizing and explaining.

This was too glib an analysis for Sigmund Freud, who by that time had abandoned his eel studies and become the star of psychoanalysis. In 1919, he published the essay "Das Unheimliche," in part as a rebuttal to Ernst Jentsch's definition of the concept. Jentsch, Freud admitted, was right to say insecurity triggers that feeling of uncanniness; for instance, when looking at a body that we're not sure is alive or dead, or when we encounter madness in another human being, or witness an epileptic fit. But not every new and strange thing is unpleasant. It takes something else, Freud claimed; another element has to be added to make the situation uncanny. What was needed was the familiar. More specifically, the uncanny is the unique unease we experience when something we think we know or understand turns out to be something else. The familiar that suddenly becomes unfamiliar. An object, a creature, a person, who is not what we first thought. A well-crafted wax figure. A stuffed animal. A rosy-cheeked corpse.

Freud turned to language to explain his thinking. "The German word *unheimlich*," he wrote, "is obviously the opposite of *heimlich*, *heimisch*, meaning 'familiar,' 'native,' 'belonging to the home'; and we are tempted to conclude that what is 'uncanny' is frightening precisely because it is *not* known and familiar." But *heimlich* is also an ambiguous word, he claimed, since it can denote that which is secret and private, that which is hidden from the world. The word contains its own opposite. And the same is, of course, true of that which is *unheimlich*; it is at once both familiar and unfamiliar.

That is how, Freud states, we should understand the unique sense of unease called *unheimlich*. It overcomes us when what we recognize contains an element of strangeness and we become unsure of what we're really looking at and what it means.

With his essay "Das Unheimliche," Sigmund Freud gave fear a psychoanalytical foundation that authors and artists have used ever since. And I would like to think the eel played at least a small part in it.

Because, after establishing the linguistic ambiguity of the concept, Freud turns to E. T. A. Hoffmann's short story "The Sandman" to demonstrate how this unique feeling of uncanniness is expressed. "The Sandman" tells the story of a young man named Nathanael, who while visiting a strange city for his studies is forced to encounter his repressed past and by extension his madness. As a child, Nathanael was told that a terrifying creature called the Sandman appears at children's bedsides in the night and steals their eyes. As a grown-up, he believes he encounters a reincarnation of the Sandman in the form of a man who sells barometers and optical instruments. And when he falls in love with a mysterious woman by the name of Olimpia, it turns out she is in fact a robot created by the barometer salesman and a professor called Spalanzani. When Nathanael eventually realizes the truth, and beholds Olimpia's lifeless body at the professor's house, her eyes lying next to her on the floor, he is overcome with madness and tries to kill Spalanzani.

The entire short story teeters on the brink of uncertainty. The narrative perspective shifts continually, nothing is truly known, things may be happening in the material world, or possibly only in Nathanael's tormented mind. To Freud, the woman who turns out to be a robot and the theft of the eyes are also central symbols at the core of the uncanny; here is an example of the uncertainty about whether a creature is alive or dead, but also the fear of being robbed of one's sight, of losing one's ability to observe and experience the world as it truly is.

But perhaps other elements of Hoffmann's story also resonated with Freud. The story is about a young German-speaking man who travels to a strange city to study. The city is never named, but both Professor Spalanzani and the barometer salesman are said to speak Italian. Furthermore, the barometer salesman doesn't just sell barometers but all kinds of optical instruments, including microscopes, the tool that is supposed to reveal the truth to the scientifically minded. Also, and this may be a coincidence, but an entertaining one, the mysterious Professor Spalanzani in "The Sandman" happens to share his name with the famous scientist Spallanzani, who in the eighteenth century traveled to Comacchio to seek the truth of the eel, in vain.

To top it off, Freud at the end of "Das Unheimliche" recounts one of his own uncanny experiences. He describes a walk in a "provincial town in Italy"; it is a hot afternoon and without quite knowing how, he ends up on a narrow street where everywhere he looks, painted women stare out of windows. He walks away, only to find himself a while later in the same place. He leaves again, but soon discovers he has circled back to the same street for a third time. Three times he has unconsciously been brought to exactly the same place, like being forced to relive the same experience again and again in a dream.

He finds it uncanny. The involuntary repetition, experiencing the exact same unwelcome scenario over and over again, kind of like standing in a dark laboratory week after week, dissecting fish after fish only to find something other than you expected. "I was glad enough to abandon my exploratory walk and get straight back to the piazza I had left a short while before."

He is, in all likeliness, writing about Trieste. He described similar, dreamlike walks in his letters to Eduard Silberstein during his 1876 visit, when he unsuccessfully tried to find the eel's testicles. The same narrow alleys and painted women watching him from the windows. It appears, then, that what came to mind when Freud himself tried to capture the unique feeling of unease and intellectual uncertainty was his frustrating and enigmatic weeks in Trieste. And surely it's not too far-fetched to think the eel played on his mind, because what has it been throughout history—in literature and art, as well as in its hidden existence just beneath the surface—if not uncanny? If not *unheimlich*?

12

To Kill an Animal

I remember Dad down by the stream, against a backdrop of moonlight and the soft rushing of the rapids, with reeds sticking out of the water like dark antennae behind him. He was standing at the bottom of the bank, just by the water's edge, clutching an eel. It was small, too small to take home and eat, really. But, as eels are prone to do, it had swallowed the hook so completely that it had disappeared down its throat; Dad was squeezing the eel, trying to jiggle the hook loose, but it kept writhing around his arm, up over his wrist, which was shiny with slime, and the hook refused to come out. Dad hissed, softly through gritted teeth: "You bastard."

As I watched, unease grew inside me. That thick slime, almost impossible to wash off, clinging to the skin of his arm and clothes like stinking glue. The eel's tiny button eyes, which seemed to stare at me but never returned my gaze. The slow movements, the body arching like a flexed muscle, twisting around its own axis until its white underbelly shimmered in the moonlight.

Dad squeezed the eel even harder, yanked at the line and tried to pry its jaw open, but it bit down hard and continued to writhe in his grasp, resisting sluggishly. Blood was dripping from the eel's mouth; Dad frowned and said, even more softly: "Bloody let go already. You bastard!" His words may have been aggressive, but his tone slowly changed, becoming gentle, pleading, almost tender. He shook his head. "No, it's not working." And I handed him the knife, the long fishing knife whose blade had been whetted so many times it was thin as a reed, and he squatted, held the eel against the ground, and firmly pushed the point of the knife through its head.

Dad liked animals a lot. All kinds of animals. He liked being in nature, by the stream or in the forest; he read books about animals and watched nature shows on television; he liked horses and dogs, and seeing an unusual wild animal made him very excited. Sometimes we went bird-watching. Just him and me with one pair of binoculars between us. We walked around aimlessly, passing the binoculars back and forth whenever we spotted a kite or a woodpecker. We didn't keep a log of the species we saw; it was never a sport to us. We just liked looking at birds.

He was fascinated by all the strange and wonderful forms life took. He told me about the bats down by the river, how they navigated using sonar. "They can't see a thing, barely as far as their own noses, but they let out these high-pitched squeaks that we can't even hear, and then they listen for the echo; when it comes bouncing back, they know straightaway if there's a mosquito or a tree trunk in front of them. It takes a fraction of a second."

Sometimes we heard rustling in the tall, wet grass and saw a frightened grass snake slip into the stream and swim away, its yellow spots like glimmering lanterns on its head. Sometimes we spotted a heron standing on the opposite bank, its neck bent like a fishing hook and its giant beak pointed down at whatever was hiding under the surface.

Dad told me about the mink that lived by the stream. A small, slender, almost entirely black creature that crept along the water's edge at night. At least that's what he said. I'd never seen it and wasn't sure Dad had either. But sometimes we would find half-eaten fish in the grass. "Must be the mink," Dad would offer.

He said they were lovely animals, but also crafty and dangerous, maybe not to us, but to the stream and the reason we visited it—the fish and the eel. "It kills for sport," he told me. He said the mink goes for mice and frogs and fish, definitely, and that it doesn't stop until it's killed everything in its path. Every time it runs into another life-form, it has to kill it. It's in its nature. It was an intruder, not just by our stream, but in the very ecosystem. It would be capable of emptying the stream of eels pretty much single-handedly. It fell to us to put things right.

So Dad built a trap. It was a simple, rectangular wooden box, maybe three feet long, with an opening at one end and some kind of trip lock meant to make sure the mink couldn't get out once it was inside. We baited the trap with a dead roach and placed it by the water's edge, at the bottom of the steep bank. Then we left it overnight while we fished for eels.

The next morning, we crept through the wet grass toward the trap as silently as we could. On the lookout for any sign of movement, listening for the sounds of the animal that was almost certain to be inside. But the trap was empty. The roach was still there, untouched. And that was how it always turned out, every time we set the trap, in many different spots along the stream. A single, reeking roach, left untouched. Not once did we see the faintest sign of the mink ever having been near it.

In time, I started doubting whether the mink was real, but more than anything I was relieved I didn't have to encounter it. Because what would we have really done if we'd caught a mink? I suppose Dad would've killed it. But how? With his bare hands? Or a knife? Would he have submerged the whole trap in the stream and drowned it? A small, slender, beautiful animal with bright eyes and soft, shiny fur. Was it right to kill an animal like that? It felt foreign, an act completely different from killing a roach or an eel.

WHAT MAKES A HUMAN DIFFERENT FROM AN ANIMAL? I KNEW NOTH-ing about that. The only thing I knew was that there was a difference and that it was irrevocable and immutable. A human is something other than an animal.

Eventually, I also came to understand that in addition to there being a difference between humans and animals, there's also a difference among different kinds of animals. That boundary was even more vague and less defined. The difference seemed to be less about the nature of the animals than about our perception of them. If you looked at an animal and saw something of yourself in it, you inevitably felt closer to it. That didn't mean killing any animal was easy, or that it should have been easy, just that there was a difference among different animals. Apparently, that was how human empathy worked. An animal looking you in the eye, you can identify with. That animal is harder to kill.

Dad liked animals a lot, but sometimes he killed them. It wasn't something he enjoyed, he took no pleasure in the violence, but he did what he thought was right. He'd been raised to believe humans have not only the upper hand and the power over other forms of life, but also a kind of responsibility. To let live or let die. It wasn't always clear how to handle this responsibility, or when it was right to do one thing or the other, but it was nevertheless a responsibility that was impossible to shirk. And it was a responsibility that required a certain level of respect. Respect for the animal, for life itself, but also respect for our responsibility for it.

He kept a shotgun at home. It sat in a closet, locked to the back; he rarely used it. Once or twice a year, he would go hunting with some men I didn't know. They left in the early hours of the morning, dressed in thick, baggy jackets and green hunting caps. Sometimes he came back holding a dead hare by its hind legs, limp and bloodstained. Sometimes he brought a couple of pheasants. But he seldom seemed to have shot them himself. He always said someone else had held the gun. He said he didn't like shooting the animals if they were standing still. A hare, flicking its ear, oblivious to the danger. A stock dove cooing in a tree. He stood there and took aim, but couldn't bring himself to pull the trigger.

But he did shoot our cat Oskar. That much I know. It was a fat and none-too-companionable black-and-white tom that spent most of the day sleeping on a sofa but slunk out the door every night, not to return until morning. Eventually he grew old and sick and tired, and one morning he was gone and I didn't really give it a second thought. Mum and Dad said he'd run away. Maybe he'd been run over by a car. I found out only much later that Dad had in fact killed him. He'd shot Oskar with his shotgun. Because he felt it was the right thing to do.

He tried to shoot Nana's cat, too. It was old and sick and tired as well; Dad took it into the woods to put it out of its misery. He managed to wrestle both the cat and the rifle into the trunk and then drove to a small clearing deep in the forest. Just as he pulled up, he spotted a covey of partridges at the edge of the trees. It was rare to get so close, and his gun was loaded and ready in the back. So he crept around the car carefully, tentatively opened the trunk with one hand and stuck the other inside to pull out the gun without letting the cat escape. But in that moment, the cat—the old and sick and tired cat—somehow got a second wind. Like a dark blur, it streaked out of the open trunk, dashing between the trees, straight toward the covey of partridges. And as the cat disappeared without a trace in the forest, the partridges took flight and raced, terrified, in the opposite direction. And Dad was left standing by the car, rifle in hand. Careless. A failure. He never saw that cat again.

MY FATHER'S VIEWS ON HUMANS AND ANIMALS, AND THE DIFFER-ence between them, had, of course, been with him since childhood. They were considered self-evident, indisputable. For me, it was never so clear cut.

Dad had grown up on a farm and had, since he was a small boy, helped keep the stable free of mice and rats. He'd caught them with his hands and killed them quickly and without fuss by throwing them hard against the stable wall. He'd seen chickens beheaded and kittens drowned. He'd been present when his father slaughtered pigs. He'd seen the pig get anesthetized and seen its throat cut and its blood drained. He'd learned how to scald its skin with boiling water so the thick bristles could be scrubbed off, and how the body was subsequently cut up, turning the living creature into chunks of meat.

As he got older, he continued to help with the slaughter, and once, he brought me with him. I might have been ten at the time. We left at the crack of dawn; when we got to his parents' the stable door was open and I caught a glimpse of the big tub full of steaming water inside, the knives and brushes on the floor, Grandad leading the pig, a large, pliant animal, up. I was excited and possibly a little bit scared; Dad must've noticed, because as we were about to head in and set to work, he turned to me and said: "Actually, I think it would be better if you went inside with Nana."

There was a graveness in his voice that surprised me, and I felt a pang of humiliation and disappointment. But when he stepped into the stable and closed the door behind him, leaving me alone in the yard, I was, more than anything, relieved.

Early one morning a few days later, we were down by the stream, pulling out our spillers. It was late summer and already warm, and the tall grass was dry and crackly. Big, heavy dragonflies hovered around our heads, and the stream flowed unusually calmly and contentedly through the rapids. I stood at the bottom of the bank, near the willow tree. Dad was about three feet away; we noticed one of our nylon lines was taut like a violin string. When I touched it, I could feel it vibrating; I grabbed it and was greeted by that familiar, undulating resistance. "It's an eel," I said out loud.

It was a fairly large one, with a dark brown back and shiny white belly; I held it firmly right behind the head and studied the fishing line disappearing into its clenched jaws. It writhed around my arm like a thick rope being tightened, all the way up to my elbow, then it suddenly let go and slapped me in the face with its tail. Thick slime covered my cheek. The smell of fish and the past and brackish seawater.

I fumbled its mouth open and saw that the line continued down its throat. The hook was buried deep; I couldn't even see the loop. I spent a few minutes jiggling the line, pulling and yanking and trying to stick my fingers far enough down its throat to grab the hook, until I heard a soft, wet crunching sound and blood started pouring out of the eel's mouth.

"It swallowed the hook," I said. "Could you take it?"

Dad bent closer and studied the eel.

"Poor little thing," he said. "It's in there good, isn't it? Now, why would you do that?"

Then he straightened up and looked at me again. "No, you take it. You can handle it."

13

Under the Sea

Despite the contradictory feeling the eel arouses, up close, in its natural habitat, it gives the impression of being fairly jovial. It rarely puts on airs. It doesn't cause a scene. It eats what its surroundings offer. It stays on the sidelines, demanding neither attention nor appreciation.

The eel is different from, for instance, the salmon, which sparkles and shimmers and makes wild dashes and daring jumps. The salmon comes off as a self-absorbed, vain fish. The eel seems more content. It doesn't make a big deal of its existence.

And thus the eel is in a more fundamental way the opposite of the salmon. Both are migrating fish, both live in both fresh and saltwater and both undergo metamorphoses, but their life cycles differ in their most essential aspect.

The salmon is a so-called anadromous fish. It breeds in freshwater and its offspring swim out to sea after about a year, spending most of their lives there. After just a few years (the salmon clearly doesn't possess the patience of the eel), the sexually mature salmon swims back up into fresh water and procreates.

The eel, for its part, makes a similar journey, but in the opposite direction. It is a so-called catadromous fish that lives its life in freshwater but breeds in saltwater.

Another, more subtle, indefinable detail also sets them apart. When the salmon wanders back up rivers and waterways, it always returns to the spot where its parents reproduced. Every salmon quite literally walks in its ancestors' footsteps. Somehow, it knows that's where it has to go. A salmon can live a free and unrestrained life in the sea, but eventually it will return to the place of its birth and join the community it was destined for. This means there are clear genetic differences among salmon populations from different waters. The salmon is, so to speak, biologically tied to its origin. It doesn't allow existential transgressions.

The eel, of course, also finds its way back to its birthplace—Sargasso, ho!—but once it reaches this vast sea, it encounters eels from all across Europe and breeds indiscriminately. Origin to an eel is not about family or biological belonging, it's simply a location. And afterward, when the tiny willow leaf drifts toward the coasts of Europe and turns into a glass eel, it chooses a waterway to wander up seemingly at random. Where it spends its adult life apparently has nothing to do with previous generations of eels; why a particular eel chooses a particular river remains a mystery. This means the genetic variation among eels in different parts of Europe is negligible. Every eel seeks its place in the world without a guide, without inheritance or heritage and existentially alone.

Perhaps the eel's fate is easier to identify with than the salmon's predestined lack of independence. And perhaps that's why the eel, with its enigmatic remoteness, remains such a fascinating creature. Because it's easier to relate to someone who has secrets, too, people who aren't immediately obvious about who they are or where they're from. The eel's secretive side is also the secretive side of humans. And seeking your place in the world on your own: Surely that is, at the end of the day, the most universal of all human experiences?

OF COURSE, I'M ANTHROPOMORPHIZING THE EEL, FORCING IT TO BE more than it is or wishes to be, which may seem somewhat dubious. Attributing human characteristics to nonhuman creatures has been a common device in, for example, literature: fairy tales and fables about anthropomorphized animals thinking, talking, and feeling, animals demonstrating morality and acting according to a set of values. It's also common in religion. Divine beings are given human form and characteristics in order to render them fathomable. The Old Norse Aesir were gods in human guise. Jesus was the son of God, but also a human. Only by being both could he represent a link between the worldly and the divine and become the savior of humankind. At heart, what's at stake is identification, the ability to see something familiar in the unfamiliar and thus comprehend it and feel closer to it. An artist painting a portrait always puts part of him- or herself in it.

But within science, anthropomorphism has never been accepted. Science claims to deal with unadulterated objectivity, the truth that reveals itself only under the microscope. It attempts to describe the world as it is, not as it seems. An eel is not a person and cannot, therefore, be likened to one. Anyone with an objective, empiricist approach to knowledge could not bring himself to speak of animals that way. To experience the world as human belongs to us alone.

But when Rachel Carson wrote about the eel, that was, nevertheless, what she did. She anthropomorphized it. She described the eel as a sentient creature with feelings, an animal with memory and reason, which could be tormented by the tribulations it was destined for or could enjoy the bright side of life. And she had her reasons for doing so. When the history of science is one day summed up, Rachel Carson will stand out as one of the people who contributed most to our understanding of not only the eel but also the vast and complex ecosystem to which it inevitably belongs.

Rachel Carson was one of the twentieth century's most prominent and influential marine biologists. She was first and foremost an expert in the ocean and its inhabitants; she wrote several groundbreaking books about marine life and eventually also became a pioneer of and icon to the burgeoning environmental movement. She was an extraordinary person in many ways.

Carson was born in May 1907 and grew up on a small farm in Springdale, Pennsylvania, a stone's throw from the mighty Allegheny River, which loops around the town. It was here, during her very first years, that she developed her lifelong interest in animals and nature. As a young child, she learned to love the forests and wetlands, the birds and the fish. The river in particular left her spellbound, as did everything in it, all the life that the water from the branched torrents brought with it on its long journey to the sea.

That being said, her professional path was by no means predetermined. Her father was a traveling salesman and her mother a housewife. The family was poor and an academic career hardly a given. But her mother, who had given up her career as a teacher when she got married, encouraged her daughter's interest in nature. She took Rachel on long walks to study plants, insects, and birds. She trained her in the art of observation and taught her how to notice details and also instilled in her a deep and loving respect for the diversity of life. As soon as Rachel Carson learned how to read and write, she started making little books, illustrated pamphlets with fact-filled stories about mice, frogs, owls, and fish. It's said she was a lonely child, with few, if any, close friends, but she never felt alone or out of place in nature. That was the world she got to know better than any other.

Eventually, she did end up going to university, at the age of eighteen, after graduating at the top of her class and after her mother sold the family china to pay her tuition. At first, she studied history, sociology, English, and French, but the central interest of her life is obvious from her very first university essay: "I love all the beautiful things of nature and the wild creatures are my friends." Two years later, when she was twenty, she had a life-changing realization. She herself described it as an epiphany. One day she suddenly realized she was supposed to dedicate her life to the ocean. The ocean was to be the focus of all her curiosity and academic talent. "I realized," she wrote later, "that my own path led to the sea—which until then I had not seen—and that my own destiny was somehow linked with the sea."

What drew Rachel Carson to the sea? The choice may seem arbitrary. She had grown up away from the coast and had never laid eyes on the ocean, never dipped her toes into its water or listened to its waves crashing against the shore. And yet it seemed inevitable. It was as though she were following a scent down a mighty river, against the current, all the way to its origin, to the sea, which is the origin of everything. That was the core of her epiphany. We all came from the sea once, and therefore anyone wishing to understand life on this planet has to first understand the sea. Much later, in her 1951 book, entitled *The Sea around Us*, she explained this insight in a way that encapsulates what sets her apart from most marine biologists, a way that is at once scientific and poetic:

> *When they went ashore the animals that took up*
> *a land life carried with them a part of the sea in*
> *their bodies, a heritage which they passed on to their*
> *children and which even today links each land animal*
> *with its origin in the ancient sea. Fish, amphibian,*
> *and reptile, warm-blooded bird and mammal—each*
> *of us carries in our veins a salty stream in which*
> *the elements sodium, potassium, and calcium are*
> *combined in almost the same proportions as in sea*
> *water. This is our inheritance from the day, untold*
> *millions of years ago, when a remote ancestor having*
> *progressed from the one-celled to the many-celled*
> *stage, first developed a circulatory system in which the*
> *fluid was merely the water of the sea.*

Thus we are all created from water, we all come from our own mysterious Sargasso Seas. "And as life itself began in the sea, so each of us begins his identical life in a miniature ocean within his mother's womb."

IN THE AUTUMN OF 1932, RACHEL CARSON HAD JUST BEGUN HER graduate studies in marine biology and kept in a corner of her laboratory a big tank of eels. She wanted to study how eels react to changes in salinity. She wanted to understand how the animal coped with the radical changes it experienced during its life cycle, how it submitted to its destiny, its long, hopeless migration and mysterious metamorphoses. She never got to finish her scientific study, but she was clearly taken with the eel. She would show off her eels to her friends and tell them about their enigmatic life cycle and long journey to the Sargasso Sea. And she would remain enamored with the eel and eventually return to it.

Her dream of an academic career came to an abrupt end, however, when Carson's father died in July 1935 and she suddenly found herself forced to financially support her mother and older sister. Continuing her at best modestly remunerated work in the laboratory was out of the question. Ambition and self-realization had to yield to duty and family loyalty. But via her contacts at the university, she was given an opportunity to earn a regular salary by indulging another long-standing interest; namely, writing. She started penning scripts for a radio series about life in the oceans. Over fifty-two episodes, each seven minutes in length, she told her listeners about many aquatic species, in a way that was both scientifically accurate and interesting to a lay audience. And her employer, the US Bureau of Fisheries, was so happy with the result that she was immediately given another assignment: to write the introduction to a pamphlet about marine life. She entitled her piece "The World of Waters," and it was a story about life in the ocean, about all the creatures lurking under the mirrorlike surface, that live their lives there, hunting or being hunted, being born, propagating and dying. It was a text that rested solidly on her academic knowledge about marine life, but it was also a creative and empathetic narrative. Her supervisor read it and declared it unsuitable for an informational pamphlet from the bureau. This was not what he'd envisioned. This was literature.

"I don't think we can use it," he said. "But submit it to *Atlantic Monthly*."

And that is how she eventually became a writer; and thus, Rachel Carson's path did in fact lead her to the sea, to the origin of everything, and her life and work would come to revolve around getting to know and understand this origin.

RACHEL CARSON'S FIRST BOOK WAS PUBLISHED IN 1941. IT WAS called *Under the Sea-Wind* and was based on her piece about the sea, which was in fact published in the *Atlantic Monthly*. She wanted to write about the sea as the vast and multifaceted environment that it is, to show at least part of what goes on in its depths, beyond the gaze and knowledge of humanity. And by doing so, she also wanted to point to something much bigger and more universal: how everything is connected. She wrote in a letter to her editor: "Each of these stories seems to me not only to challenge the imagination, but also to give us a little better perspective on human problems. They are as ageless as sun and rain, or the sea itself."

She therefore turned to an unusual literary method for a marine biologist. She used anthropomorphism, the device of fairy tales and fables. The first part of the book describes life at the water's edge; the second part is about the open sea, and the third outlines what is happening in its depths. Each part centers on a particular animal. In the first part, we meet a seabird, a black skimmer, living its life on the edge of the sea. It hunts for minnows and crabs, moving with the seasons and tides, an entire life lived as a perfectly adapted cog in a much larger and infinitely complicated ecosystem. The bird is not only given a backstory and a personality but even a name, Rynchops, derived from its Latin name, and over the course of the story, it meets a great many other animals in its unique beach environment: herons, turtles, hermit crabs, shrimp, herrings, and terns. Humans, on the other hand, are nothing but remote strangers in the background.

In the second part, we follow, in a similar fashion, a mackerel by the name of Scomber, navigating the open sea, as part of an enormous shoal, surrounded by gulls, sharks, and whales, but only ever seriously threatened when faceless humans plunge their trawls into the water.

In the third and last part of the book, we are introduced to the eel. It goes without saying that Rachel Carson couldn't have found a better representative for the compelling complexity of the sea. She explains in a letter to her publisher: "I know many people shudder at the sight of an eel. To me (and I believe to anyone who knows its story) to see an eel is something like meeting a person who has traveled to the most remote and wonderful places of the earth; in a flash I see a vivid picture of the strange places that eel has been—places which I, being merely human, can never visit."

The story begins in a small lake, Bittern Pond, at the foot of a tall hill. The lake is located almost two hundred miles from the sea, surrounded by bulrushes, cattails, and water hyacinths; two little brooks feed it. That is the scene of our introduction to our main character: "Every spring a number of small creatures come up the grassy spillway and enter Bittern Pond, having made the two-hundred-mile journey from the sea. They are curiously formed, like pieces of slender glass rods shorter than a man's finger."

Rachel Carson then homes in on a particular female eel, ten years old, which she calls Anguilla. Anguilla has lived all her life in the lake, ever since she arrived as a small glass eel. She has hidden in the reeds during the day and gone hunting at night "for like all eels, she was a lover of darkness." She has hibernated in the soft, warm mud of the lake bed, "for like all eels she was a lover of warmth." Anguilla is a creature who feels and experiences, who remembers her past and knows suffering and love. Who eventually yearns. Because when autumn comes, something is different about Anguilla. She suddenly longs to leave, a vague, wordless longing, and one dark night, she sets her course for the lake's outlet, and pushes on down rivers and brooks, the full two hundred miles to the open ocean. We follow her into the sea through obstacles and trials, toward the Sargasso. Down into the depths, toward the abysses that are the "ocean basins," far down in the shadows where the water flows, "frigid water, deliberate and inexorable as time itself."

And as Anguilla and all the other mature eels disappear, from view and human knowledge, our focus shifts to the tiny, weightless willow leaves, "the only testament that remained of the parent eels," moving in the other direction, drifting on the ocean currents in a long journey back through the ocean, over the continental shelf and toward the land that "once was sea."

Under the Sea-Wind hit American bookshops in November 1941. It was, of course, remarkably unfortunate timing. A month later, worldly affairs intervened when Japan attacked Pearl Harbor. The United States was at war, and the public's interest in fairy tales about eels, mackerels, and black skimmers was suddenly minimal. The book sold fewer than two thousand copies and was soon forgotten.

Eventually, however, it would be picked up again, published in new editions and read and loved by successive generations. Above all because it describes the sea in a way that's beautiful and fantastical, dreamlike and literary, but also always based on science. Rachel Carson's decision to anthropomorphize the animals was, of course, deliberate and in service of a purpose. She used fairy-tale devices but never went beyond the boundaries of science and fact. She didn't let the eel speak or act in a way that would be alien to the real animal. She was simply trying to imagine what reality is like for an eel, how it experiences all the hardships, metamorphoses, and migrations of the strange life cycle she also describes with scientific clarity. She explains in the foreword of the first edition, "I have spoken of a fish 'fearing' his enemies . . . not because I suppose a fish experiences fear in the same way that we do, but because I think he behaves as though he were frightened. With the fish, the response is primarily physical;

with us, primarily psychological. Yet if the behavior of the fish is to be understandable to us, we must describe it in the words that most properly belong to human psychological states."

And thus, the eel's behavior became comprehensible to us for the first time, or at least slightly more comprehensible than before. What Rachel Carson realized, and what makes her unique in the history of natural science, was that she had to be able to see part of herself in another creature in order to truly understand it. She identified with animals, and this identification gave her the ability, and the courage, to anthropomorphize them. She did something that's taboo in traditional science: she gave the eel awareness, an almost human consciousness, and thereby managed to get closer to it. She didn't do it because she believed eels posses that kind of awareness, in the strictly scientific sense, but to help us better understand what a unique and complex creature it is. To let the eel be an eel, but also something we can to some degree identify with. A mystery, but no longer a complete stranger.

SO WHAT'S THE DIFFERENCE BETWEEN AN EEL AND A HUMAN? A common definition of what makes us human is that we're aware of our own existence, and with this awareness comes a desire to affect existence. At least that's how the difference between humans and animals has been historically conceived.

In the seventeenth century, René Descartes claimed all creatures except humans should be thought of as "automata." Animals were bodies, the actions of which were nothing more than mechanical reactions. Humans, on the other hand, had something all animals lacked, a soul. The soul enabled thinking, which was in itself proof of the existence of awareness. Ergo, humans had awareness because they had a soul. Animals had no soul and therefore no awareness.

With the aid of a soul, humans were elevated above animals, but also above the passage and transience of time. The notion of a soul was and is still associated with the idea that humans are individuals. The word *individual*, in turn, means something that can't be divided, a unit that stays whole and unchanged even when everything else changes. And since the human body is unarguably changeable, as are the external conditions of a human life, there must be something else, something permanent, that makes us individuals. This something has since time immemorial been the soul.

That being said, this particular difference between animals and humans has never gone unchallenged. When Carl Linnaeus published the tenth edition of his constantly reworked *Systema Naturae* (the edition usually considered the most important because it contains the beginnings of zoological nomenclature), in 1758, it featured some controversial revisions from previous editions. This is where Linnaeus, among other things, recategorized whales from fish to mammals, and bats from birds to mammals. But this was also where he temporarily erased the line between human and animal. In this particular edition, he placed the orangutan in the same genus, *Homo*, as humans. Which meant that according to Linnaeus, the orangutan was human. That we, *Homo sapiens*, were not, after all, the only living members of our genus, that we weren't as unique as we'd always assumed.

That was a scientific mistake and it was quickly corrected, but even so, it did raise interesting questions. If the orangutan was human, did that mean the orangutan had a soul? Was it aware of its own existence? If so, what was the difference between a human and an orangutan? And if that difference was erased, what was really the difference between humans and bats or eels?

Eventually, Charles Darwin came along and robbed us of our eternal soul once and for all. The theory of evolution didn't allow for the concept of an unchangeable soul, since it posits that all life, and all parts of it, are changeable. The human became an animal among other animals. And in time, as modern science developed, the animals of the world have, conversely, become a bit more like us. They've been given if not a soul then at least awareness. We know today that animals can possess considerably more complex states of consciousness than previously thought. Research shows that most animals, including fish, can feel pain. Signs point to animals being able to experience fear, grief, parental feelings, shame, regret, gratitude, and something we might call love.

There are also animals, such as primates and crows, that can perform advanced mental tasks, that can learn to communicate and interact both with members of their own species and with others, that can imagine the future, that can decline a reward in the present in exchange for a promise of a greater reward later on. All the criteria that we have throughout history postulated as pivotal to separating humans from animals—awareness, personality, the use of tools, a concept of the future, abstract thinking, problem solving, language, play, culture, the ability to feel grief or loss, fear or love—all these criteria have been shown to be at the very least disputed, often insufficient, sometimes completely erroneous. The difference has, to some degree, in fact been erased. A crow placed in front of a mirror knows that it's looking at itself, which means it's aware of its own existence. It knows that it is, regardless of whether it can be said to know what it is.

SO THE EEL HAS AWARENESS, AT LEAST AT SOME LEVEL. BUT IS IT aware of its own existence? And if so, what does an eel feel? How does it experience its many metamorphoses, its long wait, and its migrations? Can it feel boredom? Impatience? Loneliness? What does the eel feel when that final autumn comes and its body changes, growing strong and turning silvery gray, and something profound and unfathomable urges it out into the Atlantic Ocean? Is it longing? A sense of incompleteness? A fear of death? What is it actually like to be an eel?

Rachel Carson anthropomorphized the eel in order to help us understand it better, to let us imagine the experience of the eel and better comprehend its behavior. But does that mean we really understand what the eel *itself* experiences?

That question has become increasingly key over the past few decades. The philosopher Thomas Nagel wrote a famous article in 1974 about the philosophy of mind. He entitled it "What Is It Like to Be a Bat?" And his answer to this seemingly simple question is succinct: We really can't ever know.

All animals have consciousness, Nagel posits. Consciousness is above all a state of mind. It's the subjective experience of the world, a narrative told by our senses about the things around us. But even so, a human can never fully comprehend what it's like to be a bat, or an eel, or an imagined extraterrestrial, for that matter. Our experiences as humans limit our ability to imagine the consciousness of other species.

A bat, for instance, is clearly in a completely different state of consciousness from a human. It perceives the world primarily through echoes. We know this thanks to, among others, Italian scientist Lazzaro Spallanzani, the man who aside from sharing his name with the mysterious professor in E. T. A. Hoffmann's short story "The Sandman" also unsuccessfully sought the truth about the eel's reproduction. In the early 1790s, Spallanzani conducted a number of groundbreaking experiments on bats, which, among other things, allowed him to conclude that they could fly without hindrance or collisions through completely darkened rooms. He also captured a large number of bats and removed their eyes before releasing them back into the wild. When he managed to recapture some of the blind bats a few days later, he dissected them and found freshly caught insects in their stomachs. In other words, the bats could both hunt and navigate without the use of their eyes. It followed, Spallanzani argued, that they must be using their ears.

So a bat flies over a river at night, seeing virtually nothing but sending out rapid, high-frequency noises that bounce back against the objects and creatures that surround it. The echoes of these sounds are processed and interpreted by the bat in order to build an extremely detailed picture of the world. Thanks to this ability, a bat can fly at full speed in complete darkness through the branches of a tree without crashing. It can even tell one type of moth from another by the way sound bounces off their wings. Everything the bat encounters has its own pattern of echoes, and this is how it understands its surroundings. Its perception of the world consists of a constant stream of echoes, and these echoes, of course, shape how the bat *feels* about the world.

Human consciousness is fundamentally different, and if we try to imagine what it's like to be a bat, it is that human consciousness that, according to Nagel, limits our ability to do so.

It's not enough that I try to imagine what it's like to have wings and terrible eyesight, what it's like to fly over a river at night and catch bugs with my mouth, or to imagine what it's like to emit audio signals and pick up their echo. "In so far as I can imagine this (which is not very far)," Nagel writes, "it tells me only what it would be like for me to behave as a bat behaves. But that is not the question. I want to know what it is like for a bat to be a bat. Yet if I try to imagine this, I am restricted to the resources of my own mind."

Nor is the problem, Nagel claims, limited to the relationship between humans and animals. How can, for instance, a hearing person imagine how a person who has been deaf since birth perceives the world? How can a sighted person explain a picture to a person who has always been blind?

What Thomas Nagel does reject is what's called reductionism, which is the idea that complex concepts can be explained and understood through simpler concepts. For example, that we would be able to understand the mind of another creature by studying and describing the physical or chemical processes of that creature's brain. Reductionism tries to explain big things through small things; the whole is made up of smaller components that can be explained and understood individually, and which is expected to make the whole fathomable in turn.

But it's not enough, Nagel argued. When it comes to consciousness, there are states that are completely unknown to us and will remain so, even if the human species were to survive until the end of time. Some things will always remain out of our grasp, be they about bats or eels. We can learn where these creatures come from, how they move and navigate, we can get to know them, almost as humans, but we will never fully understand what it's like to be them.

This is a logical approach to the world, and by all appearances correct. And yet it's tempting to think Rachel Carson did manage to reach a kind of understanding that shouldn't really be possible. Not through reductionism or empiricism or even science's traditional belief in truth as it appears under the microscope, but by having faith in an ability that may in fact be unique to humans: imagination.

THE FAIRY TALE GOES SOMETHING LIKE THIS: ONCE UPON A TIME, A boy caught an eel. The boy's name was Samuel Nilsson and he was eight years old. The year was 1859.

Samuel Nilsson dropped his catch, a relatively small eel, into a well on his home farm in Brantevik, in southeast Skåne, the southernmost part of Sweden. The well was then sealed with a heavy stone lid.

The eel remained there, alone in the dark, kept alive by the occasional worm and insect that would fall into the water, cut off from the world and robbed not only of the sea, the sky, and the stars, but also the meaning of its existence: the journey home, back to the Sargasso Sea, the thing that would make its life complete. And the eel lived on while everything around it disappeared. The eel lived on while at the end of the nineteenth century its contemporaries grew strong and shiny and set their course for the Sargasso to spawn and die. It lived on while Samuel Nilsson grew up and old and eventually died. It lived on while Samuel Nilsson's children did the same. And his grandchildren and great-grandchildren.

The eel lived for so long it eventually became famous. People traveled from far and wide to look down the well and maybe catch a glimpse of it. It became a living link to the past. An eel robbed of life that had gotten its revenge by cheating death. Perhaps it was even immortal?

Calling it a fairy tale is really neither right nor fair, though. That there really was an eel in the well in Brantevik is indisputable. That it had been there a long time is by all appearances equally true. Only the bit about Samuel Nilsson is slightly difficult to verify. Exactly how long the Brantevik eel had lived in its well can't be established beyond doubt.

Nevertheless, some have tried. In 2009, the Swedish nature television program *Mitt i naturen* visited the farm in Brantevik. At that point, the eel was, according to legend, one hundred and fifty years old, and by documenting its existence, the crew wanted to shift at least some aspect of it from the world of myth to that of reality.

It was one of Swedish nature television's most dramatic moments. The TV team managed to heave the big, square stone lid aside and look down into the well, which was no more than fifteen feet deep and lined with large stones. There was, of course, no sign of the eel. They set up a pump and drained the well of water. Still no sign of the eel. The host, Martin Emtenäs, climbed down and searched the cracks between the stones as water trickled back in. Still no sign of the eel.

They were just about to put the big stone lid back when they suddenly spotted movement in the murky water at the bottom of the well; Emtenäs climbed back down to check what it might be.

The eel, the mysterious Brantevik Eel, which they finally managed to pull out, was a strange creature. It was small (twenty-one inches long), thin, and pale, but with abnormally large eyes. While all other parts of it had shrunk to adapt to life in the cramped, dark well, its eyes had grown several times larger than a normal eel's—as though it was trying to compensate for the light it had lost. Slithering through the grass next to the well, it looked like a visitor from another world. So tragically marked by a life of darkness and solitude. So odd and alien once it was pulled up into the light to join the rest of us.

"It's perfectly possible the myth of the Brantevik Eel is true," Emtenäs mused afterward. Perhaps it really was one hundred and fifty years old. After it had lived for a century and a half in those conditions, the TV crew probably felt it would be high-handed to disturb the order that had let the eel cheat death for so long. After measuring and examining the eel, they dropped it back into the well, back into the darkness where it seemed intent on surviving us all.

The Brantevik Eel survived for a few more years before finally giving up. In August 2014, the owner of the well discovered it was dead. Its remains were shipped to a laboratory in Stockholm, where it was hoped the number of rings on its otolith, a kind of calcareous organ of the inner ear, would establish its age once and for all. Unfortunately, no otolith was ever found; perhaps the tiny crystalline structure had disappeared when the body decomposed. The sediment at the bottom of the well was dug up and sifted through, but the otolith wasn't there either. Somehow, the eel managed to cheat humanity one last time, even if it had grown too weary to cheat death.

REGARDLESS OF WHICH ASPECTS OF THE LEGEND OF THE BRANTE- vik Eel are true, it's a fact that eels can live for a very long time. The oldest eel whose age has more or less been verified was caught in Helsingborg in 1863 by a twelve-year-old boy named Fritz Netzler. The eel was a couple of years old at the time, thin, and no more than fifteen inches long. It had arrived from its long journey from the Sargasso Sea, transformed from glass eel to yellow eel, and had wandered into Öresund and up a waterway called Hälsobäcken, which at the time ran straight through a park in central Helsingborg. There, before the eel had made it more than a few hundred yards up the waterway, Fritz Netzler caught it. He named the eel Putte and kept it in a small tank in the apartment in Helsingborg where he lived. The eel grew older, but not much bigger. The years passed and the eel remained in a juvenile state, thin and just over fifteen inches long.

Putte was about twenty when Fritz Netzler's father, whose name was also Fritz and who was a doctor, died, and for a while the eel and its captor were separated. Putte and his tank moved from family to family in Helsingborg. He might also have lived in Lund for a while.

He was nearly forty when in 1899, he moved back in with Fritz Netzler Jr., who by then was a man and a doctor just like his father. Putte was still thin and just over fifteen inches long, and after so many years in tiny tanks in dark flats, his eyes had grown disproportionately large, just like the Brantevik Eel's. It's said Putte would eat out of Fritz's hand. Meat or fish; his favorite was calf liver cut into small pieces.

Eventually, the eel outlived its captor. Putte was nearing his seventieth birthday when Fritz Netzler Jr. died in 1929, and after a few years with yet another family, he was finally donated to the Helsingborg Museum in 1939. That's where Putte eventually passed away, at ostensibly eighty-eighty years old, in 1948.

Putte was stuffed and is today kept in storage at the museum. According to its catalog, the item consists of "Putte the eel in tank with lid, containing eel in fluid and rocks." The tank is twenty inches long. Putte himself, in taxidermized form, is just under fifteen.

And so Putte the eel likely lived for almost ninety years and was still, in human terms, more or less a teenager. Because, like the Brantevik Eel, Putte wasn't just an eel that remained remarkably small; he never underwent the last metamorphosis that would have turned him into a sexually mature silver eel. Which points to another mysterious aspect of the eel question: How does the eel know to initiate its various transformations? How does the eel know when life is coming to an end and the Sargasso Sea is beckoning? What kind of voice lets it know it's time to leave?

It can't just be random. Because apparently the eel is capable of suspending its own aging, no matter how long it lives for. When circumstances require it, its final metamorphosis is postponed indefinitely. If the eel isn't free to go to the Sargasso Sea, it won't undergo the final metamorphosis, won't turn into a silver eel, and won't become sexually mature. Instead, it waits, patiently, for decades, until the opportunity presents itself or it runs out of strength. When life doesn't turn out the way it was supposed to, an eel can put everything on hold, and postpone dying almost indefinitely.

When a scientific study in Ireland in the 1980s caught a large number of sexually mature silver eels, it was discovered that the age of the fish—which were on their way to the Sargasso Sea and thus in the final stage of life—varied significantly. The youngest was only eight and the oldest fifty-seven. They were all in the same developmental phase, the same relative age, if you will, and yet the oldest was seven times older than the youngest.

You have to ask yourself: How does a creature like that perceive time?

To humans, the experience of time is inevitably tied to the process of aging, and aging follows a fairly predictable chronological trajectory. Humans don't undergo metamorphoses in the technical sense; we change but remain the same. Overall health can, of course, vary among individuals; we can suffer illness or injury, but generally speaking, we know roughly when to expect a new phase; our biological clock is not particularly flexible; we know when we are younger and when we grow older.

The eel, by contrast, becomes something else each time it transforms, and each stage of its life cycle can be drawn out or condensed depending on where it is and what the circumstances are. Its aging seems tied to something other than time.

Does a creature like the eel even experience time as a process, or more like a state? Does it, simply put, have a different way of measuring time? Oceanic time, perhaps?

Rachel Carson claimed that in the sea, deep down where the eel spawns and dies, time moves differently from how it does for us. Down there, time has somehow outlived its usefulness and is irrelevant to the experience of reality. Down there, our regular chronological measurements don't exist. There is neither night nor day, winter nor summer; everything unfolds at its own pace. Rachel Carson wrote her book *Under the Sea-Wind* about the abyss underneath the Sargasso Sea, where "change comes slow, where the passing of the years has no meaning, nor the swift succession of meaning." And she wrote *The Sea around Us* about sailing across the open ocean on a starry night, gazing toward the distant horizon and feeling that neither time nor space is finite: "And then, as never on land, he knows the truth that his world is a water world, a planet dominated by its covering mantle of ocean, in which the continents are but transient intrusions of land above the surface of the all-encircling sea."

The oldest creatures we've found so far all came from the sea. Ming the clam, a so-called ocean quahog caught off the coast of Iceland in 2006, turned out to be at least five hundred and seven years old. Scientists estimated its year of birth to be 1499, a few years after Columbus made it to North America and during the time of the Ming dynasty in China. Who knows how long it could have lived if the scientists in their efforts to establish its age hadn't also accidentally killed it. In the Pacific Ocean, east of China, there are organisms called glass sponges, which, it's been shown, have the ability to live for over eleven thousand years. At the bottom of the sea, where the earth's orbit and the rising and setting of the sun are meaningless, aging seems to follow a different law. If there really is something eternal, or nearly eternal, the ocean is where we'll find it.

EELS MAY NOT BE IMMORTAL, BUT THEY ALMOST ARE, AND IF WE AL- low ourselves to anthropomorphize them slightly, we must in- evitably ask ourselves how they handle having so much time. Most people would say there's nothing worse than boredom. Ennui and waiting are fiendishly hard to endure, and time is never as present and persistent as when we're bored. One shudders at the mere thought of a hundred and fifty years at the bottom of a dark well, alone and practically in sensory de- privation. When there are no events or experiences to distract us from time, it becomes a monster, something unbearable.

I imagine a hundred and fifty years alone in the dark as an endless, sleepless night. The kind of night when you can feel each second being added to the one before, like a slow, interminable jigsaw puzzle. I try to imagine the impatience of a night like that, being so utterly aware of the passing of time and yet so utterly unable to speed it up in the slightest.

To the eel, things are, it would seem, different. An animal probably doesn't experience tedium the same way humans do. An animal doesn't have a concrete notion of time, of sec- onds turning into minutes and years and whole lifetimes. Per- haps boredom doesn't make eels impatient.

But there's a different kind of impatience, which may be relevant. It's the one we feel when we are forced to endure lack of fulfillment. The impatience at being stopped from doing what you set out to do.

That's what I think about when I think about the Brantevik Eel. Even if it lived to a hundred and fifty, no matter how long it managed to postpone death, there wasn't enough time for it to make its predestined journey and complete its existence. It overcame every obstacle, survived everyone around it; it managed to draw out its long and hopeless life—from birth to passing—for a century and a half. Yet even so, it never got to go home to the Sargasso Sea. Circumstances trapped it in a life of endless waiting.

From this we can learn that time is unreliable company and that no matter how slowly the seconds tick by, life is over in the blink of an eye: we are born with a home and a heritage and we do everything we can to free ourselves from this fate, and maybe we even succeed, but soon enough, we realize we have no choice but to travel back to where we came from, and if we can't get there, we're never really finished, and there we are, in the light of our sudden epiphany, feeling like we've lived our whole lives at the bottom of a dark well, with no idea who we really are, and then suddenly, one day, it's too late.

14

Setting an Eel Trap

We lived in a white brick house—my mother, father, older sister, younger sister, and me. We had a garage, a lawn, fruit trees, and a greenhouse in which Mom and Dad grew tomatoes. We all had our own rooms, and there was a bathroom with a tub, a decent-size kitchen, and a living room with paintings on the walls where no one ever spent any time. We had a TV room with a large sofa. We had a basement with a laundry room and a boiler room. We had a garden with potatoes, carrots, and strawberries, and a compost pile where you could dig for worms. We had a Ping-Pong table, a loom, and an extra freezer, and a still for making moonshine, which every other month or so bubbled away in the shower, sending a strong smell of mash throughout the house. We had an apple tree and a plum tree, which together formed a perfect soccer goal. We had a sandbox and a conservatory with a plastic roof that pattered like rifle fire when it rained. We lived on a street where all the houses had been built at the same time. Our neighbors were butchers, pig farmers, janitors, and truck drivers, and there were children everywhere. We were completely unremarkable. We were amazingly unremarkable. That was the only thing that made us special.

I understood early on that the life Mom and Dad had made for themselves had not been a given. They were both from somewhere else and had ended up where they'd ended up because people like them had been swept along in a process that in three short decades had changed almost everything. It wasn't individual class mobility, it was collective. Three decades of social reform in Sweden had moved the working class, at least parts of it, from laborer's cottages and cramped apartments to their own houses, complete with garages, fruit trees, and greenhouses. It had been a mighty movement, like an ocean current.

Dad was born in the summer of 1947. His mother, my grandmother, was twenty years old at the time and had already been working for more than six years. After seven years in school, she had her Confirmation and then, at the age of fourteen, started working as a maid. The morning after her Confirmation, she rode her bicycle to her first job. She had bought the bike on credit, paying it off in monthly ten-kronor installments. Her salary was twenty-five kronor a month.

She lived with her parents and five siblings. Her parents were agricultural contract workers who were paid in kind with food rather than money: a whitewashed form of slavery. The family lived in a typical contract worker's cottage. Three rooms: a kitchen, a bedroom in which all eight members of the family slept—two to a bed—and a parlor no one was allowed in during the day. Outhouse, wood-burning stove, and drafty windows. A violent father. They were people without possessions, and even after the contract worker system was abolished in 1945, they stayed on in the house, living and working much like before. Contract workers knew their place. As did the children of contract workers.

My grandmother was beautiful in a simple, unpretentious way; she smiled often and had shy eyes with a touch of melancholy about them. She worked as a maid in about ten different households during her teens. Doing dishes, dusting, and so on, from seven in the morning until seven at night. She had Sundays and one afternoon a week off. She slept alone in a maid's room and she was unhappy—unhappy being a maid, unhappy living as a stranger in other people's homes, unhappy with the scoldings and contempt and submission. She was constantly homesick, for her sisters and brothers and childhood.

Right before my father was born, my grandmother moved back in with her parents and found work at the rubber factory in town. She preferred working in the factory to being a maid, but she was also a single parent to a small child. She was given two months of parental leave and then had to go back to work. Her parents and younger sisters were in charge of my father during the day.

He was seven when he and Nana moved to the farm by the stream.

It was a tenant farm, owned by the church, with pigs and fields and a garden full of flowers that my grandmother cared for. Dad was put to work on the farm from the start, but he also liked boxing and using a slingshot. He ran across the fields to the stream and learned to swim just above the rapids. He went to school and was interested in history and science but eventually dropped out. He started working, transporting pigs for the abattoir. He did his military service and met mom and got a job as a paver, which he kept until the end of his days.

While Dad was growing up, Sweden introduced universal child support, income support, and occupational pension. Income taxes had been individualized. Healthcare, maternity care, childcare, and elderly care had all been expanded. Wealth had been redistributed. Two guaranteed weeks of vacation had expanded to four. Society and the state had taken over large sections of the social safety net from families. In other words, it had become possible for a road paver and a day care worker mom, my parents, to live a life that was different in every way from the lives previous generations of the working class had known.

Nothing about my parents' life was a given, of course. But nor was it chance. Strong forces had been involved. They had been willow leaves in a mighty current. They had traveled across an ocean without really moving at all.

Dad was twenty and mom seventeen when they had my big sister. Just a few years later, they took out a loan from the bank and built the white brick house.

ONE DAY, MY DAD PLACED A LONG, NARROW, STRANGE-LOOKING object made of metal hoops and mesh on the lawn in front of the house.

"It's an eel trap," Dad told me. "I bought it."

I don't know who he bought it from; either way, it wasn't new; there were several large holes in the mesh, which we mended with sewing thread, but there was something awe-inspiring about it. It was about fifteen feet long, considerably wider at one end and tapering toward a point at the other, and it had two mesh wings by the opening that could be extended out to either side, making it at least ten feet wide. I pictured it on the bottom of the stream, catching everything carried into it by the current. It would be full to the brim with fish. This was something other than setting spillers. This was something that upset the balance of power. With this trap, we would no longer be temporary, unobtrusive guests in the constant cycle of life and activity in the stream; we would be almost omnipotent. It was as though we could now intervene in the fundamental order of things.

We had dinner and Dad pushed some snus up under his lip and then we were on our way down to the stream while there was still light. We skidded down the slope and drove along the wide tracks, parking by the willow tree. It had been raining for days and the water level was high; the stream was at least a few feet wider than usual and bursting its banks in places, forming small pools of stagnant water, out of which solitary blades of grass protruded.

Our boat was moored next to the willow tree, tearing at its chain like a trapped animal. Dad stood motionless, studying the murky water rushing by both faster and with more force than usual. "I'll be damned, the water has risen," he said and spat in the grass. "All right, let's give it a try anyway."

We'd brought the sledgehammer, two long poles, and one shorter one; we put them and the trap in the boat and pushed off.

"Want me to row?" I asked.

"No, I'll do it," he replied. "You set it up."

He rowed some way into the stream, turned, and started struggling against the current, away from the rapids. The crutches squealed when he heaved at the oars. The current pushed back at every stroke, lifting the prow straight up. He muttered and cursed and leaned his whole body back every time he pulled. After about a hundred yards, he stuck the oars almost straight down and braced with his arms, trying to keep the boat still. It lurched from side to side as if trying to tear free. Dad pumped the oars to parry the movements.

"Take the long one and bang it into the bottom," Dad said, nodding impatiently toward the side. Fumbling, I found the pole and plunged the sharp end into the water, pushing it into the muddy streambed as hard as I could. The boat dashed about as though it were trying to buck me, but I managed to reach the sledgehammer and get in some half-decent blows. Brown, dirty water splashed my face.

We were wet and filthy by the time I'd finally managed to bang down both long poles and tie the wings at the trap's opening to them. Dad's face was shiny and he was breathing heavily. He raised the oars and let the boat glide along for a few feet so I could set up the shorter pole as well and tie the tapered end to it. The trap spread out before us, hidden in the murky water, with its opening in the middle of the stream and its mesh bag like a secret room beneath the surface.

Dad pulled the oars out with a sigh and let the boat float along at will. He spat in the water and looked at the two poles sticking up like the masts of a sinking ship.

"This should bloody well get us some eels."

That night, I fell asleep with images of eels flashing before my eyes. Tons of eels, flashing yellow and brown, crawling around my feet. They were gaping and glaring and gasping for air, struggling to climb up my legs like creepers climbing toward the light. Their eyes were like black buttons.

The next morning, the water had already subsided a little. Dad was holding the oars, studying the stream. The current seemed to have slowed, the water had cleared, and he didn't have to try quite as hard to turn the boat against the current and row toward the trap.

But we could tell from a distance something was amiss. One of the long poles stood slanted in the water, the other was missing entirely. The whole trap had been pulled along and overturned so the wide opening was pointing downstream instead of upstream, secured now only to the short pole.

"Damn it!" Dad said.

He rowed up to the short pole. The trap was swaying this way and that; I yanked the pole up and hauled in the cold, wet mesh, which was covered in dark green plants. The water soaked my trousers, and my hand grew numb; Dad put the oars up and took the trap in silence, tossing branches and large clumps of shiny seaweed overboard, folding the mesh into a pile between us.

That's when I spotted it. At the very apex of the narrow end, partially hidden by seaweed, was an eel, writhing sluggishly from side to side. It was the size of a blindworm, just over seven inches long, thin and with tiny black dots for eyes, and I thought that it shouldn't have had a problem getting out through the mesh.

It goes without saying it was too small to keep, but we put it in the bucket anyway.

"I want to bring it home," I said.

"What for?" Dad asked. "It's too small to eat. Better leave it to grow."

"I could keep it in the tank, the one in the basement," I said.

Dad smiled and shook his head. "An eel as a pet . . ."

When we got home, I put the tank in my room. It was small, maybe a foot and half across; I poured sand into it, added a big rock, and filled it with water. I dropped the eel into the tank; it sunk to the bottom almost without moving and settled behind the rock.

I never named it. Over the weeks that followed, the eel just lay there behind the rock, and I sat next to the tank, staring at it through the glass, waiting for it to move, for something to happen, to suddenly see something behind its seemingly dead black eyes. I tried to feed it, dropping small bugs and worms into the water, but it didn't react. Just lay behind the rock as though hibernating, as though time had ceased to exist.

I tried to imagine what it saw when it looked out through the glass, what it felt. Was it scared? Was it playing possum? Did it think the world had ended when it was ripped from its usual environment? Could it imagine an existence other than the one it had now?

After a month, I still hadn't seen the eel move. It was lying dead still behind the rock. Its tiny gills pulsating gently on the sides of its head, the only sign of life. The water was getting murky. It reeked of decay.

"It's not eating," I told Dad. "It's going to starve to death."

"Oh, it'll eat when it needs to, I'd wager."

"But it's not moving either. I think it's dying."

A few days later, Dad came to my room and checked the tank. He looked at the filthy water and eel behind the rock, frowned, and shook his head.

"No, this is pointless."

That night, we returned to the stream and I carried the bucket down the bank from the car; by the willow tree, I put it down and picked up the eel. It felt cold and lifeless; I lowered my hand into the water and released it. At first, we were both motionless. Then the eel moved. Its body undulated slowly from side to side, and with gentle motions, it swam back down into the dark and disappeared.

15

The Long Journey Home

An eel, silvery and fat, swims out to the ocean, setting off on its final journey back to the Sargasso Sea. How does it know where to go? How does it find its way?

When it comes to the eel, we can allow ourselves to ask banal questions, simply because the banal questions don't always have immediate answers. We can also allow ourselves to welcome this. We should be glad that knowledge has its limits. This response isn't just a defense mechanism; it's also a way for us to understand the fact that the world is an incomprehensible place. There is something compelling about the mysterious.

Because what does it really mean when we say we know the eel procreates in the Sargasso Sea? It means we have good reason to believe this, given Johannes Schmidt spent eighteen years sailing back and forth across the Atlantic, catching tiny, transparent willow leaves. We choose to put our faith in Schmidt's work, in his observations and conclusions. We believe mature silver eels swim all the way back to the Sargasso Sea to spawn, that it's the only place they breed and that none of them leave there alive. We believe it because everything points to its being true and because no one has offered any plausible alternatives. We can even go as far as saying we know that's how it is. "We know now the destination sought," Johannes Schmidt wrote. After all his years on the open sea, he must have felt he had the right to substitute belief for knowledge.

And yet, in this case, any knowledge comes with qualifications. What we rely on when we say we know where the eel procreates isn't just observations but also a number of assumptions. And for a person who wants to know for sure, that's obviously a problem. If you want to be categorical about it, which the scientifically minded tend to want to be, knowledge is not a matter of degrees; it's binary. You either know or you don't. Science is much stricter than, for instance, philosophy or psychoanalysis in that regard. Sciences like biology and zoology have on fairly solid grounds clung to the conviction that data need to be empirical and that knowledge requires observation.

To some extent, that's the ghost of Aristotle still haunting us. All knowledge must spring from experience. Reality has to be described as it appears to our senses. Only what we've seen can be said to be true. It's an interpretation of how humans acquire knowledge about the world that has survived because it's logical, but also because it carries within it a promise. Before we know it, we have only faith, but the person with patience is always rewarded eventually. The truth will appear under the microscope.

When we say we know the eel procreates in the Sargasso Sea, there are still some essential objections to that statement: (1) No human has ever seen two eels mate. (2) No one has ever seen a mature eel in the Sargasso Sea.

That means the eel question remains unanswered; the truth has not yet appeared under the microscope. This uncertainty clearly acts as a driving force and a gravitational pull for eel enthusiasts. The mystery is there to be solved, questions await their answers, but at the same time the riddle is what sparks and perpetuates interest. For centuries, people who have viewed the eel question as a problem to solve have at the same time clung almost lovingly to the enigma of it.

When Rachel Carson wrote about the eel in her fairy tale–like nature book *Under the Sea-Wind*, she lingered on the mysterious and unexplained. Being a natural scientist, she could have been frustrated by not knowing, but the opposite seems to have been true. Rachel Carson seems to have been drawn to the uncertainty. She approached the eel and nature not just as a scientist but as a human being.

For instance, about the silver eel's long journey to the Sargasso Sea, she wrote: "As long as the tide ebbed, eels were leaving the marshes and running out to sea. Thousands passed the lighthouse that night, on the first lap of a far sea journey. . . . And as they passed through the surf and out to sea, so they also passed from human sight and almost from human knowledge."

Aristotle, Francesco Redi, Carl Linnaeus, Carlo Mondini, Giovanni Battista Grassi, Sigmund Freud, or Johannes Schmidt might have objected—perhaps they would have been unable to accept that a creature can in fact leave the realm of human knowledge—but to Rachel Carson, there seems to have been something simple and beautiful about the idea of the eels vanishing into the unknown. A creature that actively seeks to avoid human knowledge. As if that's the way it should be. "The record of the eels' journey to their spawning place is hidden in the deep sea," she wrote. "No one can trace the path of the eels." To her, the eel question, the enduring mystery, seems to have appeared to be preordained and eternal. As though it were a riddle beyond our human comprehension. Like infinity or death.

Tom Crick, the history teacher and narrator of Graham Swift's novel *Waterland*, clings to the same feeling of a kind of fated inexplicability when he expounds on the eel: "Curiosity will never be content. Even today, when we know so much, curiosity has not unraveled the riddle of the birth and sex life of the eel. Perhaps there are things, like many others, destined never to be learnt before the world comes to its end. Or perhaps—but here I speculate, here my own curiosity leads me by the nose—the world is so arranged that when all things are learnt, when curiosity is exhausted (so, long live curiosity), that is when the world shall have come to its end. But even if we learn how, and what, and where, and when, will we ever know why? Why, why?"

IN SPITE OF ALL OBSERVATIONS AND ATTEMPTS TO UNDERSTAND (until the end of time), there is thus still a lacuna in the story of the eel. We know silver eels leave in the autumn, when the eel darkness descends, usually between October and December. The tiny willow leaves, the leptocephalus larvae, appear in the

Sargasso Sea in the spring; the smallest specimens usually between February and May. Which should mean breeding happens around this time. Which in turn gives us a time frame for the eel's journey. It has at most six months to get there.

Yet even so, it's something of a mystery why the eel sets its course for the Sargasso Sea and nowhere else. Lots of animals migrate for breeding purposes, but few undertake a journey as long and difficult as the eel, and few are as stubbornly fixated on one single place thousands of miles away, and few do it just once before dying.

There are theories claiming only the Sargasso Sea has the right temperature and salinity for the eels' propagation. It's also a fact that eels have been around so long the continents have moved; the first eels likely had a much shorter distance to travel. But as the landmasses of our planet have changed, drifting apart inch by inch over the years, the eel has refused to adapt. It still needs to return to its birthplace, to the exact location it once came from.

More than anything, it's still a mystery how the eel gets there. What route does it take? How does it find its way and how does it get there on time? How can an eel make it almost five thousand miles from the rivers and waterways of Europe across a deep ocean to the other side of the Atlantic in just a few months?

In 2016, a European research team published a report on the most extensive study ever of the European eel's journey toward the Sargasso Sea. Over five years, a total of seven hundred silver eels had been tagged with electronic transmitters and released from different locations in Sweden, France, Germany, and Ireland.

As the eels turned west and the transmitters eventually fell off and floated to the surface, loaded with information, the researchers could form a picture of what their journey actually looks like.

At least that was the idea, but as is so often the case where eels are concerned, things didn't turn out as planned. Of the seven hundred transmitters, only two hundred and six yielded any information at all. And of those two hundred and six eels, only eighty-seven got far enough into the sea for their information to reveal anything useful about what their journey had been like.

But data from eighty-seven silver eels' journeys toward the Sargasso Sea is still far more than we had before, and the results revealed a lot about what a complex and difficult process this yearly migration really is. The first finding was that the eels swam both day and night and seemed to employ a deliberate strategy to avoid danger. During the day, they moved through the darker and much colder water at a depth of about three thousand feet. At night, under the cover of darkness, they rose up toward the warmer water nearer the surface. Even so, a large proportion of the eels disappeared during the earliest stages of the journey, falling prey to sharks and other predators.

What the researchers could also see was that not all eels are in a hurry. In theory, the journey to the Sargasso Sea is plausible. Experiments have shown that an eel swimming at normal speed moves slightly farther than half its length every second, and a silver eel on its way to the Sargasso Sea, which no longer hunts or eats or lets any of life's distractions slow it down, can swim without stopping for at least six months using nothing but its fat reserves as fuel. If you draw a line on a map, from any given place in Europe to the Sargasso Sea, and calculate how fast it would need to swim in order to arrive by May at the latest, the eel's journey is certainly possible. Very long and difficult, but possible.

Among the eels in the study there were, however, many that didn't seem to realize what was actually required of them, or how little time they had. A few impressive individuals did cover an average of thirty-one miles a day, but others managed only two.

The eels also chose wildly disparate routes. Clearly, many roads lead to the Sargasso Sea. The majority of the eels released on the Swedish west coast, for example, chose a northerly route, up through the Norwegian Sea and then west across the northeast Atlantic. They all chose roughly the same path, apart from a single eel, which after reaching the Atlantic suddenly veered east and disappeared without a trace outside Trondheim, Norway.

The eels released in the Celtic Sea south of Ireland and in the French Bay of Biscay, on the other hand, headed south before turning west. One of them meandered about west of Morocco for more than nine months before making it all the way to the Azores.

The eels released off the German Baltic coast took different routes. Some followed the Swedish eels, setting their sights on the Norwegian Sea. Others headed south through the English Channel. But none of them reached the Atlantic.

The eels released from the French Mediterranean coast swam, predictably, west toward Gibraltar, but only three of them managed to get through the straits and into the Atlantic.

At first, the results looked random, to say the least. The eels' movements traced strange patterns on the map, as though someone had tried to draw a maze blindfolded, or as though nothing was predetermined and every journey was the first. But at least one thing was made unambiguously clear: the majority of eels never make it to their spawning grounds. The long journey back to their birthplace remains for most of them a thwarted aspiration.

That may seem like a bleak outcome, both for the eels and for the scientific study. Not one of the seven hundred silver eels released could be tracked all the way back to the Sargasso Sea. It's impossible to say if any of them reached it. Sooner or later, they disappeared into the depths, leaving the realm of human knowledge while their electronic transmitters floated up to the surface.

Nevertheless, the research team managed to draw some new and fairly remarkable conclusions from their observations. Their initial finding was that the eels' migration is likely more complex than previously thought, but that it could be explained—at least in part. Because from the observations that at first seemed random and unpredictable, a pattern eventually appeared. Firstly, it was clear that the eel rarely takes the shortest route from its starting point to its goal. Its journey isn't like the journeys of birds or airplanes. Nevertheless, all of Europe's eels seem to rendezvous somewhere around the Azores, about halfway through their journey, and continue west toward the Sargasso Sea from there in much closer formation. If the journey starts in uncertainty and slight confusion, it becomes more deliberate as it progresses.

The researchers also discovered something else that complicates our understanding of the eels' migration. When old specimens of leptocephalus larva caught in the Sargasso Sea were reexamined and compared for size and growth rate, they showed that the eel's spawning season probably starts earlier than previously thought, possibly as early as December. That would mean breeding commences around the same time the last silver eels set off from the coasts of Europe, which only serves to make the question of how they get there on time even more vexing.

But the explanation, the researchers claimed, must, of course, be that all eels don't make it across the Atlantic in time for the next breeding season. For some, the long journey back to the Sargasso Sea can take much longer. Perhaps eels simply adjust their speed and route according to their abilities. While some swim as fast as they can in order to reach the Sargasso Sea in early spring, some take a considerably more leisurely approach and wait for the next breeding season instead. While an eel setting off from Ireland, for instance, can travel west in an almost straight line and get there by spring, an eel coming from the Baltic Sea might aim to arrive in December, more than a year after it first set off. That would not only explain the differences in the behavior observed but also lend some kind of logic and relevance to what at first seemed random. Maybe eels are, quite simply, individuals, who not only have different abilities but also different means and methods of reaching their goal. Maybe they're all aiming for the same destination, but no two journeys back to the origin are exactly the same.

AND THUS, ONE QUESTION REMAINS, AND IT IS ONE THAT APPLIES TO both eels and humans: How do they know which route will take them back to where they came from? How do they find their way back home?

That the eel has special abilities that make it skilled at navigating great distances has long been known. It's well established, for example, that it has a phenomenal sense of smell. According to the German eel expert Friedrich-Wilhelm Tesch, who wrote the standard reference work *The Eel* in the 1970s, the eel's olfactory sensitivity is on par with a dog's. Put one drop of rosewater in Lake Constance, Tesch claimed, and an eel can smell it. It's likely that eels use smell in some way during their journey across the Atlantic, either to locate the Sargasso Sea itself or at least one another. It's also likely that the eel is sensitive to changes in temperature and salinity and that these might offer clues as to which way to go. Some scientists believe the eel's well-developed magnetic sense constitutes its primary navigational tool. Much like bees and migrating birds, it can feel the earth's magnetic field and is thus guided toward a certain destination.

We know what that destination is. And somehow, the eels know it, too. They know where they're going, even if the routes they choose can be both meandering and unpredictable. But how they know is one of the mysteries still surrounding the eel question, one of the enigmas even scientists cherish.

Rachel Carson, for her part, described the eel's inherited knowledge about its origin as something more than an instinct. In *Under the Sea-Wind*, she writes about how the fully grown and sexually mature eels one autumn suddenly feel a "vague longing for a warm, dark place," and how these eels, who have lived their long lives "beyond all reminders of the sea," in lakes and rivers, now set off into the unfamiliar open ocean, finding there something familiar, something they recognize, a sense of belonging "in the large and strange rhythms of a great water which each had known in the beginning of life."

Do they remember where they came from and where they're going now? Do they remember their very first journey across the Atlantic as tiny, transparent willow leaves? No, perhaps not in a human, conscious sense, not according to our definition of memory. But when the European research team who followed the more or less successful attempts of seven hundred eels to reach the Sargasso Sea tried to explain how the eels find their way back to their birthplace, they still described the experience as a kind of memory. It seemed, they wrote, as though "eels follow olfactory cues originating in the spawning area or that eels navigate using oceanic cues imprinted or learned during the leptocephalus phase."

Because what their study revealed more than anything was that the farther the eels got, the more they seemed to end up following a predetermined route. Simply put, they seemed to follow the Gulf Stream and the North Atlantic Drift, but in the opposite direction. As though a memory, a map, had been ingrained in them when they made the journey from the Sargasso Sea to Europe as tiny, transparent willow leaves, and as though that memory had survived in the eels, remaining constant through all their metamorphoses, for ten, twenty, thirty, or fifty years, until one day it was time to make that same journey in reverse, straight toward the mighty ocean current that had once carried them helplessly to Europe.

AND SO THE SILVER EEL FINALLY COMES HOME TO ITS BIRTHPLACE, its Sargasso Sea, and at the same time, it disappears out of sight and our realm of knowledge. No one has ever seen an eel in the Sargasso Sea.

Some have tried, however. After Johannes Schmidt's years-long expeditions in the early twentieth century, it would be a while before anyone set off for the Sargasso Sea to look for the eel again, possibly because Schmidt's work was so persuasive, but perhaps even more likely because it was so discouraging. But the past few decades have seen an increase in research traffic to the Sargasso Sea, expeditions manned by some of the most prominent eel experts in the world. They've gone to seek deeper knowledge of the eel's migrations and reproduction, to test existing theories by verifying or disproving them, but also to find what no one has yet been able to: a living eel in the Sargasso Sea.

The German marine biologist Friedrich-Wilhelm Tesch went on a major expedition with two German ships in 1979, the eventual result of which was the much-cited article "The Sargasso Sea Expedition, 1979." The expedition took place in the spring and roved across large parts of the eel's supposed spawning area. Tesch was able to employ his nets and trawls in the exact location where breeding was thought to occur; like Schmidt, he caught large numbers of tiny leptocephalus larvae, but other than that, he found no sign of the presence of eels. For example, seven thousand fish eggs were collected, but closer examination revealed that not a single one came from an eel. It goes without saying that researchers didn't see any mature breeding eels either.

The American marine biologist James McCleave, who for more than thirty years has been one of the world's leading eel experts, went on his very first marine expedition together with none other than Friedrich-Wilhelm Tesch in 1974 and undertook his first journey to the Sargasso Sea in 1981. Since then, he and his team have returned seven more times, using a range of sophisticated methods to try to catch at least a glimpse of an eel. McCleave has posited a theory according to which areas where different bodies of water of different temperatures meet—so-called front regions—provide eels with exactly the right conditions for procreation. It is in such locations that he has caught the smallest specimens of leptocephalus larvae, and it is also where he has most zealously looked for adult eels. James McCleave has sailed back and forth across these regions, with ships equipped with advanced acoustic instruments designed to pick up echoes from breeding eels in the deep. And he has, in fact, recorded echoes very likely produced by living, breeding eels; each time he has tried to catch them, however, his nets have come up empty.

During one expedition, together with a fellow marine biologist, Gail Wippelhauser, McCleave employed almost malicious cunning to lure the shy eels out of the depths. Their team had caught a hundred fully grown female American eels and injected them with hormones to induce sexual maturity. The plan was to bring these females on their expedition and place them in cages fastened to floating buoys in the middle of a front region in the Sargasso Sea. The females were intended as bait, to attract males who had swum there to spawn, and thus force them out of hiding.

But the eels were reluctant participants. The scientists kept the mature females in a laboratory and were about to drive them down to the docks in Miami ahead of departure, but before the ship had even cast off, the majority of the eels had died. By the time the expedition arrived in the Sargasso Sea, only five of the one hundred female eels were still alive.

Regardless, the five surviving eels were placed in cages and tied to the buoys, and McCleave and Wippelhauser took turns monitoring the movements of the buoys around the clock with the help of radar. But inexplicably, they managed to lose them. Eels and cages and buoys disappeared without a trace and were never seen again.

During another expedition, which Gail Wippelhauser undertook without James McCleave, the acoustic instruments picked up echoes from what was believed to be a large group of breeding eels; the researchers threw at it everything they had, lowering no fewer than six nets into the water. And yet there was no sign of any eels.

Another strange detail is, of course, that it's not only living eels that have proved elusive in the Sargasso Sea. No one has ever spotted a dead one either, whether in the form of a corpse or as the victim of a larger predator. Swordfish and sharks have been caught with silver eels in their stomachs, but never anywhere close to the Sargasso Sea. A sperm whale was once caught off the Azores with an eel in its stomach that was on its way to spawn, but the Azores are pretty far from the Sargasso Sea. Once eels reach their breeding ground, they universally manage to avoid human detection in both life and death.

It should be said that there is no consensus on how significant it would really be to find a mature eel in the Sargasso

Sea. Some scientists feel it's beside the point, since we already know that's where the eels are going. Others claim our knowledge of the eel's life cycle can't be considered complete until someone has observed an eel at its spawning ground. To these scientists, the elusive eel is something of a scientific holy grail.

In the past few decades, some researchers, such as James McCleave, have started asking another difficult question: If we can't track all silver eels back to their birthplace, and in fact not even a single one, can we really be completely certain the eel breeds only in the Sargasso Sea? Granted, it took Johannes Schmidt almost twenty years to find the smallest of the tiny willow leaves there, but he had searched only a fraction of the world's oceans. Schmidt himself wrote in 1922 that until all the seas have been trawled for eel larvae, it would be impossible to say for certain where the eel breeds, or at least where *all* eels breed. And virtually all eel expeditions since, including James McCleave's, have focused on the already familiar region of the Sargasso Sea. Perhaps some eels go elsewhere entirely? It's unlikely, but how can we know for certain?

Moreover, the Sargasso Sea is very large. Is it one big breeding ground, or are there several separate breeding grounds within its borders? Do the American and European eels breed in exactly the same area, or do they prefer different locations? Some scientists, Friedrich-Wilhelm Tesch among them, have claimed that the American eel breeds in the western part of the Sargasso Sea while the European one stays farther east, but that the areas are partially overlapping. Others argue the collected leptocephalus larvae do not support such conclusions. All we know for sure is that when the tiny, transparent willow leaves leave the Sargasso Sea, European and American ones are intermingled, drifting helplessly along in the mighty ocean currents, while their parents appear to remain, die, and decompose.

HENCE, TO THIS DAY, THE WORLD'S LEADING ZOOLOGISTS AND MA-rine biologists, the people who are most intimately familiar with the eel, are forced to qualify their reports and results with reservations. "We believe," they're obliged to say. "The data indicate . . ."; "It can be assumed that . . ." By patiently rejecting less likely scenarios, they are slowly moving toward a probability that in turn closes in on truth.

It can, for example, be assumed that what's true of one of our eel's closest cousins, the Japanese eel, is also true of the European eel. And when it comes to the Japanese eel, some of the classic aspects of the eel question are, in fact, slightly less enigmatic.

The Japanese eel, *Anguilla japonica*, looks essentially like its European counterpart. Its life cycle is also very similar. It hatches in the sea and drifts toward the coast as a willow leaf. It turns into a glass eel and wanders up waterways in Japan, China, Korea, and Taiwan. It becomes a yellow eel and lives out its life in fresh water before many years later turning into a silver eel and wandering back out into the sea to spawn and die. It's a very popular fish for cooking, particularly in Japan, and it has long played an important role in East Asian culture and mythology, among other things as a symbol of fertility.

When it comes to the question of procreation—where and how it happens—the Japanese eel was long an even bigger mystery than the European one. Scientists were able to pinpoint its spawning ground only in 1991. Employing the same method and dedication as Johannes Schmidt, though not taking quite as long, the Japanese marine biologist Katsumi Tsukamoto sailed around the sea with nets and instruments, searching for increasingly minute leptocephalus larvae. One autumn evening in 1991, he finally managed to find specimens that were only days, or perhaps hours, old. It was far out in the Pacific Ocean, just west of the Mariana Islands.

After this discovery, it wasn't long before an even more sensational discovery was made. In the autumn of 2008, a research team from the Atmosphere and Ocean Research Institute in Tokyo actually managed to catch fully grown Japanese eels in exactly the area west of the Mariana Islands where the findings situated the breeding area. One male and two females were caught. All three had already spawned and were in bad shape. They died shortly thereafter. But this meant the Asian version of the holy grail of science had at long last been found.

But what did that mean? According to at least one member of the expedition, Michael Miller, nothing, really. It didn't prove anything we didn't already know. We already know approximately where they breed. But we still don't know exactly where, how they get there, or how many of them are successful. We still haven't seen them procreate. We don't know why. Why, why?

MYSTERIES HAVE AN ALLURE OF THEIR OWN, BUT THERE ARE SOME things that suggest the timeless eel question will eventually be answered. Not only have silver eels been found after breeding in the Pacific, but researchers there have also pulled off what no one has managed with the European or American eel. They have successful bred the Japanese eel, *Anguilla japonica*, in captivity. As early as 1973, scientists working at the University of Hokkaido were able to extract eggs from sexually mature female eels, inseminate them artificially, and have them hatch and become larvae. The future of the threatened eel was not their primary concern; the venture had rather narrower economic motivations. The eel is vastly popular on Japanese dinner tables and the subject of a multimillion dollar industry. If it could be farmed, the way salmon is, for instance, it would mean a lot more eel at a fraction of the cost. Consequently, the market is prepared to invest large sums in research that could make farming possible.

Unsurprisingly, the eel has not, however, proved particularly cooperative. The sensational artificially produced little willow leaves at the University of Hokkaido barely had time to hatch and register the lack of ocean currents in their tank before they died. The leptocephalus larvae simply refused to eat. It didn't matter what the Japanese researchers tried to tempt the transparent little creatures with. The willow leaves went on hunger strike and invariably perished.

For years after that, and over many generations of artificially created but all equally short-lived leptocephalus larvae, Japanese scientists dedicated themselves to finding out how to keep newly hatched eel larvae alive. What do they eat? No one knew. Their feeding habits had never been observed in the wild. A range of foods were offered. Plankton, roe from other fish, microscopic rotifers, parts of octopuses, jellyfish,

shrimp, and clams. The tiny larvae stubbornly refused suste-
nance in each successive attempt and predictably died soon
after hatching.

It took the scientists close to thirty years to come up with
a meal the larvae could stomach. It consisted of a powder
made of freeze-dried shark eggs; armed with this, they managed to keep a handful of larvae alive for all of eighteen days
in 2001. It was a sensational new record, but they were still,
of course, very far from finding the answer to how to coax the
transparent willow leaves into transforming into fully grown,
edible eels in captivity.

Furthermore, the eels continued to be difficult in other
ways. Even though the researchers were now able to make
them eat—the prescribed diet was refined over time until at
least some specimens survived into the glass eel stage—most
still died within a few days of hatching. Only 4 percent of the
larvae lasted for fifty days, and only 1 percent for a hundred.
The number that reached the size necessary to turn into glass
eels was almost zero.

Moreover, the laboratory eels behaved differently than
their peers in the sea. The captured females produced significantly fewer eggs in captivity than in the wild. It also soon
became clear that all the eels hatched in the laboratory were
male. No one knew why, but to remedy it, glass eels were injected with estrogen to artificially produce females. In 2010,
Japanese scientists succeeded for the first time in completing
the life cycle of the eels when they produced eggs, and in time
leptocephalus larvae, from eels that had themselves been created in the laboratory. The eels were also given hormones to

make them grow faster, which lead to severe deformities in their offspring: willow leaves that didn't look anything like the ones caught in the sea, their heads strangely misshapen, and the animals themselves unable to swim. It was as though the eel were refusing to let anyone else control its creation. As though its existence was its own business.

As of this writing, scientists are working hard to find the correct methods, if they even exist, to farm eels, which would be important not only to the Japanese eel industry but also, by extension, to the survival of the eel globally. They are nowhere near succeeding. But every year brings new technologies, new scientific insights and innovations, and for anyone interested in understanding the eel, there is—all the obvious problems notwithstanding—reason for hope. Perhaps some kind of tracking device will be developed in the not-too-distant future that's small and light enough to follow a silver eel all the way to its breeding grounds in the Sargasso Sea. Perhaps that will allow us to pinpoint more precisely where on the map reproduction takes place, and perhaps once enough eels have been tracked, we can confirm or reject the idea of multiple breeding grounds. Perhaps by then, we will also have a better understanding of what stops or impedes the eel on its journey back to its birthplace. Perhaps we can even do something about it. Perhaps European and American researchers will, like their Japanese colleagues, manage to fertilize eggs from European and American eels and hatch them in captivity. Perhaps one day, these cultivated eels will survive and grow big and healthy enough to be eaten. Or, of course, to be released into the wild.

A scientifically minded optimist would say it's just a matter of time. With a focused will and enough time, science will find a way to solve every riddle. The eel question has endured in various guises over millennia, but experience tells us we will find the answer, sooner or later. We just need enough time.

The problem, though, is that time is about to run out.

16

Becoming a Fool

I remember Nana on the lawn. With her head slightly bowed and her arms raised in front of her. She was holding a branch broken off the apple tree next to her. It was the first time I saw a dowsing rod.

She slowly walked across the grass, away from the tree, turned left and then right, searchingly, as though every step was a step into the unknown. Her eyes were vacant, as though she wasn't even aware that we were standing there watching.

Suddenly, she stopped; her arms twitched and were pulled down toward the grass. The rod seemed to tug at her, hard and violently, as though trying to wrest free of her grasp. And Nana looked up and laughed and said: "I can't explain it. It's not me doing it. I'm not even moving."

Dad shook his head, walked over to her and grabbed the tree branch with one hand. Then they held it together while they slowly walked around, side by side, in a circle on the grass, like a slow, peculiar dance; when they got back to that spot, they stopped, and Nana's arms were once again pulled violently downward. Dad looked up and laughed, too, while the branch was still moving.

"I can barely hold it," Dad said.

When he let go, Nana stopped moving. She held the branch up in front of her and looked at it in wonder.

"I can't explain it. But I can feel it. It's pulling all by itself."

"I just don't get it," Dad said.

One night by the stream, Dad put the bucket with our fishing gear down and broke a Y-shaped branch off the willow tree. He pulled off all the twigs and leaves and held it up in front of him.

"Should we try?"

I nodded, a little nervous, and watched him walk off slowly, in his orange waders and big, bulky wellies. He walked carefully and slightly bowlegged along the stream, away from me through the wet, somewhat unyielding grass. When he turned around and looked at me, he was a silhouette in the evening sun; I saw him holding the branch out in front of him, tentatively and almost reluctantly, as though it were leading him toward something he didn't quite know whether he wanted to meet. He walked all the way back to me without anything happening, and when he reached me, he stopped, tossed the branch aside, and shook his head.

"No, nothing. I guess I don't have the gift."

What neither Dad nor I knew then was that there's a simple explanation as to why a dowsing rod moves. The explanation has, in fact, been known for more than a hundred and fifty years. Numerous scientific experiments have been conducted to test the dowsing rod's ability to locate things such as water, oil, or metal underground. Virtually all of them have shown that it simply doesn't work. A tree branch is incapable of conveying any information whatsoever about what exists or doesn't exist underground.

And yet, it moves. Sometimes, evidently, without the person holding it deliberately trying to affect it. The explanation is what's called the ideomotor phenomenon. What happens is that a type of minute muscle movement is executed without the conscious intent of the person in question. Rather than deliberate acts, these movements are the expression of an idea, a feeling, or a perception. It's sometimes called the Carpenter effect, after the English physiologist William B. Carpenter, who first described the phenomenon in 1852, and it's the exact same phenomenon that, for example, moves the planchette on a Ouija board.

In other words, a person holding a dowsing rod unconsciously causes it to strike the ground through tiny, barely perceptible movements. But for it to work, the person has to have an idea or preconceived notion, an unconscious will leading him or her to a certain spot. Not necessarily the right spot, whether the goal is to find water or metals, but to a specific spot nonetheless. What does the unconscious find there, when the branch tugs our hands down toward the ground? Why do the muscles move in one spot but not others?

The ideomotoric effect cannot explain this, of course. Maybe it depends on our subtle sensory impressions. Maybe we subconsciously read our surroundings and come to conclusions we don't even understand ourselves. Either way, we're making these same unconscious decisions continuously.

Perhaps, after all, it's just chance that tells us when it is time to move a muscle. When it is time to stay, or when it is time to leave.

NANA BELIEVED IN GOD.

"He's big," she'd tell me. "Much bigger than anyone you can imagine."

"Is he bigger than grandad?" I asked.

"Much bigger!"

She didn't go to church, but she believed in God. In Jesus and the Immaculate Conception and the resurrection. And a life after death in which she would meet her mother and father and eventually her older siblings and her husband. And in the end, her son. She believed in gnomes, too. She'd seen one when she was about fifteen and working as a maid. She'd been walking home late one night along a tree-lined gravel road and suddenly, he'd been walking there next to her on the verge. A gnome. Dressed in gray. Barely three feet tall. She'd been with a friend who'd seen him, too. For a while, the little creature had walked beside them, then he'd vanished.

I wasn't a believer. I went to the children's group in our local church but was kicked out because I couldn't sit still, and when we attended church with school, I raised my hand and asked the priest: "Who on earth made all this up?"

Dad wasn't a believer either. He'd been to school and learned about the Swedish kings of yore and the gospel, but he had a hard time with authority. He believed in neither gnomes nor God.

It was only where the eel was concerned that we had our doubts.

Once, when we checked our spillers in the morning, we found we'd caught only one lousy eel. Granted, it was fairly large, almost two pounds, grayish-yellow and broad headed. We put it in a bucket of water in the garage as usual.

That afternoon, I went out to change the water and discovered the eel was gone. The bucket was tall and white and filled with water to a point about ten inches below the rim; the eel had been hovering near the bottom, pumping its gills the last time I checked on it. Now it was gone. The bucket was still upright and full of water, but no eel.

I didn't know what to think. At first, I figured it had managed to heave itself out of captivity and slither away. But the garage door had been closed and there was no sign of it; the eel had seemingly vanished without a trace. Had Dad cleaned it already? Without me? It didn't sound likely, but he wasn't home and wasn't expected back all day. Maybe he'd taken care of the eel before he left after all.

When Dad got back that night, I met him at the car.

"Did you take the eel?"

"The eel? It's in the bucket, isn't it?"

"No, it's gone. Someone must've taken it."

We went into the garage and stood there for a minute, staring at the empty bucket. Dad confirmed the eel really wasn't there.

"But I don't think anyone would take an eel," he said. "It seems an odd thing to steal. I think it escaped. It must be around here somewhere."

We searched the entire garage. It was dirty and full of stuff. Wooden boards, ladders, tools, plastic crates, shovels, pitchforks, rakes, buckets, potato crates, and fishing gear. We moved everything, examining every nook and cranny.

We finally found the eel in a corner, behind a pair of wellies. It lay completely still, covered in dust and gravel. I picked it up; its body was cold and limp, its skin dry and rough from the gravel. It drooped like a dirty sock in my hand; its eyes were flat and lifeless.

It was clearly dead. It had been out of the water for at least five or six hours. Maybe more.

"Put it in the bucket; I'll see to it later," Dad said.

I dropped it into the water and stood there studying it for a while. At first, it floated on the surface, its pale belly facing up. Then it suddenly turned over. Its body writhed and its head swung from side to side and slowly, slowly, it started swimming around the bucket, its gills opening and closing.

I'd seen this before. Early one morning by the stream, while it was still dark out, we'd trudged down the bank to a spiller set on a small ledge, maybe three feet above the water. On the line running out over the edge dangled an eel. Not in the water but in the air, with its head almost level with the spiller and the tip of its tail an inch or two above the surface of the stream.

I'd heard about eels catching their prey and then spinning their bodies around their own axes in a violent spiral. This eel had apparently spun so violently it had wrapped itself in the line and then kept going until it was lifted out of the water and left dangling in midair.

It hung there quietly, its head lolloping to one side. I picked it up. Several yards of thick nylon line was wrapped tightly around the eel; it had bitten into its skin, leaving bloody stripes along its entire body, as though it had been lashed. I gently untangled the line and held the eel in my hand; it felt limp and heavy and dead. Then I put it into the bucket and watched it float belly up for ten seconds, twenty seconds, before it slowly turned over and started swimming along the inside.

THERE ARE CIRCUMSTANCES THAT FORCE YOU TO CHOOSE WHAT TO believe, and for as long as I can remember, I've been the kind of person who chooses to believe what people consider verifiable, science over religion, the rational over the transcendental. But the eel makes that difficult. For anyone who has seen an eel die and then come back to life, rationality isn't enough. Almost everything can be explained; we can discuss different processes of oxygenation and metabolism or the eel's protective secretion or its highly adapted gills. But on the other hand, I've seen it with my own eyes. I'm a witness. An eel can die and live once again.

"They're odd, eels," Dad would say. And he always seemed mildly delighted when he said it. As though he needed the mystery. As though it filled some kind of emptiness in him. And I let it sway me, too. I decided that you find what you want to believe in when you need it. We needed the eel. The two of us wouldn't have been the same without it.

It was only much later, when I read the Bible, that I realized that this is exactly how faith arises. Having faith is to approach the mystery, that which lies beyond language and perception. Faith requires you to give up part of your logic and rationality. Paul wrote as much in his first letter to the Corinthians: "Your faith might not rest in the wisdom of men but in the power of God." Put differently, a believer must let go of intellectual thought, must let himself be convinced, not by rational argument or natural science or the truth that reveals itself under the microscope, but by feeling alone. "If any one among you thinks that he is wise in this age, let him become a fool that he may become wise," Paul wrote. Anyone who seeks faith must dare to become a fool.

Only a fool can believe in miracles. There's something both terrifying and tempting about it. When Jesus walks on water, his apostles, who are sitting in a boat, are frightened at first. They think he's a ghost. But Jesus tells them: "Take heart, it is I; have no fear," and Peter dares to step out onto the water to meet him. That first step, when Peter lifts his foot over the boat's railing and puts it down on the water's surface, is the beginning of everything. The familiar meets the unfamiliar. Something he thought he understood turns out to be something else entirely. And he chooses to believe it. When Jesus reaches the boat, the apostles all fall to their knees and say: "Truly, you are the son of God."

When they're out sailing on Lake Galilee and a storm blows up, the apostles are frightened and wake Jesus, who is sleeping in the stern. Jesus rebukes the wind and says: "Peace! Be still!" and the wind ceases immediately. "Why are you afraid? Have you no faith?" he says reproachfully, almost mockingly.

I've never been able to bring myself to believe in the miracles of any religion, but I can understand why someone would want to swap fear for conviction. I can understand that a person coming across something unfamiliar or frightening chooses the miracle over ongoing uncertainty. It's a human thing to do. Having faith is giving yourself over to something. To what can be explained only through similes.

And the promise of the Christian faith, what awaits anyone brave enough to become a fool, is the biggest of all promises: "He who believes in me, though he die, yet shall he live, and whoever lives and believes in me shall never die."

Jesus promises his followers eternal life, which is why the most important miracle is the resurrection. That Jesus dies and is raised is the heart of the Christian message. Without it, faith becomes meaningless. Faith can't be only about this life; it has to transcend it. Paul writes in his letter to the Corinthians: "If Christ has not been raised, then our preaching is in vain and your faith is in vain."

Only a fool would believe in the resurrection, but I've sometimes wished I were a fool, and I think Dad wished for the same thing. Because what is resurrection? If taken literally it means a person (or an eel) can die and then live again. But Paul also talks about something else in his letter to the Corinthians. "The last enemy to be destroyed is death," he writes. Death is inevitable, but there are, according to Paul, ways to handle it. Further on, Paul talks about change, about how death isn't an ending but rather a kind of metamorphosis: "We shall all be changed, in a moment, in the twinkling of an eye, at the last trumpet. For the trumpet will sound, and the dead will be raised imperishable, and we shall be changed."

So a person (or an eel) can die and then be transformed in the blink of an eye and come back in imperishable form. No, that's not true. That's a simile. But a simile can carry within it its own truth, of course. You don't have to believe the miracle to believe the meaning of the miracle. There are many ways to be a fool. And you don't have to believe in the Gospel (or the eel) in a literal sense to believe what is at the heart of their message: Those who die stay with us in some form.

Nana believed in God, but Dad and I didn't. That being said, much later, when Nana was dying, I sat by her side and she cried and said, "I will always be with you." And I obviously believed her. I didn't need to believe in God to believe that.

And that is, at the end of the day, what Jesus promises his followers. "I am with you always, to the close of the age," he says when he reveals himself to his apostles, three days after his death.

And that is, of course, what we hope for when we believe. Whether in God or an eel.

17

The Eel on the Brink of Extinction

The last enemy to be destroyed is death. That's true not only for people of faith, but also for those who prefer knowledge. It's certainly true for all the people still trying to understand the eel.

Because the eel is dying out, and at an increasing rate. There are data that suggest that the eel population began to shrink as early as the eighteenth century, which is to say around the same time science first took an earnest interest in the creature. More reliable data showing a decline in eel numbers are available from the 1950s at least. And during the past few decades, the problem seems to have accelerated significantly. According to most research reports, the situation today is more or less catastrophic. The eel is dying, and not just in the expected way, as the natural end to a long life full of changes. It's becoming extinct. We are losing it.

This is the latest and most urgent eel question: Why is it disappearing?

It may be appropriate as a starting point to place the extinction of the eel within a larger context. Life is changeable; that's the first law of evolution. Life is also transient; that's the first law of life. But what's happening now with the eel, as with so many other species, is far beyond the normal progression of evolution and life, in terms of both character and extent.

Rachel Carson was one of the first to realize this. Her final book, and the one that she'll forever be remembered for, was *Silent Spring*. It was published in 1962 and is one of the most influential works ever written about humanity's ability to destroy what it claims to love. *Silent Spring* is about the devastating use of DDT and other synthetic pesticides, about how the thoughtless spraying of fields and forests kills not only insects but also all other forms of life: birds, fish, mammals, and in the end, humans. Through a combination of thorough scientific research and her inimitably beautiful and visceral language, Carson was able to both illustrate the extent of the problem and describe what it actually meant in practice.

What she foresaw was a time when life is no longer seen or heard around us, simply because it has disappeared from the world we perceive, because it has ceased to exist. She foresaw a silent time, springs without the whirring of insects or singing of birds, without fish jumping in rivers or bats flitting through the moonlight at night. She saw an ongoing destruction of large swaths of the life we were so used to having around us, and she knew why it was happening: "As man proceeds toward his announced goal of the conquest of nature, he has written a depressing record of destruction, directed not only against the earth he inhabits but against the life that shares it with him."

By identifying with the animals, with something beyond herself, Rachel Carson was able to arrive at a greater understanding of what was happening. From that sprung a feeling of desperation that eventually grew into courage and a conviction that it was her right, even her duty, to bear witness to what she knew. And that time was short. In June 1963, while *Silent Spring* sent ripples across the world, she appeared before the US Senate's subcommittee on environmental hazards; she began her statement by saying: "The problem you have chosen to explore is one that must be resolved in our time. I feel strongly that a beginning must be made on it now, in this session of Congress." Her eagerness and haste were not only rhetorical. She was dying herself. By the time *Silent Spring* was published, she had been diagnosed with breast cancer, and when she testified before the Senate subcommittee, the cancer had spread to her liver. She knew it was her last chance to turn her conviction into action—and she was successful, at least as far as the devastating pesticides were concerned. The use of DDT in agriculture was banned in the United States in 1972, largely thanks to *Silent Spring*'s enormous impact. But by then, Rachel Carson was dead. She passed away in April 1964, at the age of fifty-six. Her legacy will always be the attention she drew early on to the threat that has now become a widespread concern.

SEVERAL TIMES DURING THE MORE THAN THREE BILLION YEARS THAT life has existed on this planet, changes have taken place that have been so far reaching and drastic one could say they were tantamount to a kind of metamorphosis, each altering the very composition of life on earth. Five times, these changes have been so all encompassing, they've been given their own category. These five periods are usually called the five mass-extinction events.

The first of the mass-extinction events started about 450 million years ago, during the tail end of the Ordovician period, when life was still more or less confined to the oceans. Due to a cooling climate, which was in turn a consequence of continental drift, approximately 60 to 70 percent of all species became extinct over a period of about ten million years.

The second mass-extinction event was also caused by devastating global cooling, about 364 million years ago; by its end, 70 percent of all living species had been wiped out.

The third mass-extinction event was the deadliest. It occurred in the transition between the Permian and Triassic periods, approximately 250 million years ago, and killed off more than 95 percent of all species. There is no consensus on the cause, but the most likely answer is that a confluence of events led to dramatic climate change.

The fourth mass-extinction event took place over a relatively long time between the Triassic and Jurassic periods, about 200 million years ago, and saw the demise of up to 75 percent of all species.

The fifth mass-extinction event is the most famous. Sixty-five million years ago, a meteor is thought to have struck the Yucatán Peninsula; the impact was at least one contributing factor to the extinction of the dinosaurs, along with 75 percent of the rest of the world's species.

The flora and fauna of our planet have undergone more metamorphoses than that, some almost as comprehensive, but relative to the long history of life, mass extinctions are nevertheless a very rare phenomenon. Species die, animals and plants come and go, but the time frame of this process is usually so long it doesn't fundamentally disturb the order of things. That is the normal way of life: occasional goodbyes, not holocausts.

And yet many researchers are positing that what we are experiencing now is not the normal way of things, that we are, in fact, living the sixth mass-extinction event. In August 2008, the American biologists David Wake and Vance Vredenburg wrote an article entitled "Are We in the Midst of the Sixth Mass Extinction?" It was published in the scientific journal *Proceedings of the National Academy of Sciences*, and even though the authors were not the first to ask this question, their answers were so persuasive that the threat no longer seemed hypothetical but rather highly probable.

Wake and Vredenburg focused specifically on amphibians and salamanders and were able to show that yes, some form of mass extinction was unquestionably already underway. Of the earth's circa 6,300 known amphibian species, at least a third were already endangered, and this development showed every sign of getting rapidly worse.

One of the people who read the article was the science journalist Elizabeth Kolbert. Her book *The Sixth Extinction* was published in 2014 and summarized what we know about the potential extinction event happening right now. About a third of all corals are at threat of extinction; so are a third of all sharks, a quarter of all mammals, a fifth of all reptiles, and a sixth of all birds. This extinction event may not turn out to be as far reaching as any of the big five, but the threat is nevertheless so great, and accelerating so rapidly, that it's not out of the realm of possibility. If things carry on like this, there's much to suggest that the number of species on our planet will be halved in just one hundred years.

That is exceptionally fast—previous mass extinctions took place over millions of years; now we are talking about centuries—but what makes the current extinction event truly unique is that for the first time in history, there's a living perpetrator. The culprit is not a celestial body, or continental drift or volcanic eruptions; it's a creature. One of the many species inhabiting this planet has conquered it, and in so doing has caused the massive destruction of the habitats of all other species. It has managed to change not just the surface of the earth but its atmosphere, too. No other species has ever come close to exercising that kind of impact on life. On different forms of life. On all life.

"If Wake and Vredenburg were correct," Elizabeth Kolbert writes, "then those of us alive today not only are witnessing one of the rarest events in life's history, we are also causing it."

BUT WHY IS THE EEL IN PARTICULAR DYING? WHAT ARE THE SPECIFIC circumstances that have made this seemingly timeless survivor unable to carry on? To start, the question is accompanied by a theoretical problem. As we know, asking why can never be the first step in tackling a scientific problem. One has to start at the beginning. First, we establish that something's happening: Is the eel dying? Then we observe it and explain what is happening: How is the eel dying? Only once that has been done can we begin to approach the question of why.

And when it comes to the question of the eel's extinction, this approach has turned out to be a little bit complicated.

The name of the organization coordinating much of the work on environmental protection and biological diversity around the globe, and which has over a thousand member organizations, is the International Union for Conservation of Nature, or IUCN. Among other things, the IUCN compiles the so-called Red List, an inventory of animals and plants that is regularly updated to identify which species are considered threatened around the world. The explicit aim of the Red List is to create a "universally accepted system of classification of species at high risk of extinction globally." In other words, the IUCN's criteria serve as a kind of international standard, a scientifically tested assessment of how life in its different forms is doing.

On the Red List, each species is assessed according to established criteria and rated on a scale ranging from the most heartening ("least concern") through "near threatened," "vulnerable," "endangered," "critically endangered," and "extinct in the wild," to the final and irrevocable declaration of "extinct." And since it is an objectively and methodically compiled inventory of all known life on earth, it provides information on how everything from algae to ringworms and humans are faring.

Humans are doing well. The most recent IUCN assessment of *Homo sapiens*, from 2008, says the following: "Listed as Least Concern as the species is very widely distributed, adaptable, currently increasing." It is also noted that "humans have the widest distribution of any terrestrial mammal species, inhabiting every continent on earth (although there are no permanent settlements on Antarctica). A small group of humans has been introduced to space, where they inhabit the International Space Station." At present, according to the IUCN assessment, "no conservation measures are required." *Homo sapiens* is thriving.

The eel, *Anguilla anguilla*, on the other hand, is in trouble. Or, at least there's good reason to think it is. It's what we are led to believe. It goes without saying that since it's the eel we're dealing with, we can't claim to know for certain. As is so often the case, our knowledge comes with caveats. Because it turns out the eel doesn't quite fit the criteria normally used by the IUCN for its assessments. Firstly, our inability to determine the exact size of the total population is a problem. Population size is, naturally, the first criterion for determining the level of threat to a species. But according to IUCN's reports, population size should be determined by the number of "reproductive individuals," which is to say the number of fully grown, sexually mature specimens. That means, the IUCN writes, that ideally, the criterion would be applied to "mature eels at their spawning grounds." In other words, a headcount of silver eels in the Sargasso Sea would be necessary. However, since no one has managed to find so much as one silver eel in the Sargasso Sea after more than a hundred years of trying, it is obviously impossible. The eel won't let itself be mapped that way. It avoids even those who would help it.

What could potentially be done is a count of how many mature silver eels set off from the coasts of Europe toward the spawning grounds. But here, too, data is scarce; eels have a habit of disappearing into the dark depths of the ocean very quickly. The observations that have been made, however, suggest that the number of migrating silver eels has plummeted by at least 50 percent in the past forty-five years.

The third-best alternative, which is what the IUCN primarily bases its assessment on, is quite simply to start at the other end and assess what emerges as the result of the eels' secretive rendezvous in the Sargasso Sea—what Rachel Carson called "the only testament that remained of the parent eels." In other words, the number of glass eels that turn up in Europe in the spring. A lot more is known about this, and it's these data that suggest the situation is absolutely catastrophic. All reliable counts indicate the number of newly arrived glass eels in Europe today is only about 5 percent of what it was at the end of the 1970s. For every hundred transparent little glass rods swimming upstream every year when I was a boy, at most a handful make that same journey today.

This is the basis for IUCN's decision to categorize the European eel, *Anguilla anguilla*, as critically endangered. Which, according to the official definition, means it's "facing an extremely high risk of extinction in the wild." The situation is not only catastrophic but also acute. The eel could really disappear, in the foreseeable future, and not just from our sight and our realm of knowledge, but from our world.

SO THIS IS THE FINAL QUESTION: WHY IS THE EEL DYING? AND THE final answer is not surprising, given that this is the eel we're talking about: It's hard to say. It's the same problem that everyone attempting to understand the eel has been confronted with: The answer eludes us. We don't know for certain. We know parts, but not the whole. We are, to some extent, forced to rely on faith.

There are several explanations as to why eels are in trouble, and science can confirm them all, but no one knows for sure if they're the only causes, or even the most pivotal ones. As long as there are unanswered questions about the life cycle of the eel, we can't say for certain why the eel is dying. As long as we're uncertain exactly how the eel procreates or how it navigates, we can't say what's preventing it from doing those things. In order to save it, we have to understand it. This is what most research on the state of the eel emphasizes nowadays: In order to help the eel, we need to know more about it. We need more knowledge and more studies, and time is short.

And thus, we arrive at the great paradox: The mysteriousness of the eel has suddenly become its greatest enemy. If it is to survive, humans have to coax it out of the shadows and find answers to the remaining questions. And that will, of course, come at a cost. Because throughout history, there have been people who have embraced this mysteriousness, who have been drawn to it and have chosen to cling to it. People who, like Graham Swift, or his storyteller Tom Crick, want to believe that a world where everything's explained is a world that has come to an end.

It is, if you will, a classic catch-22: Those of us who want to protect the eel in order to preserve something genuinely mysterious and enigmatic in a world of enlightenment will, in some ways, lose no matter how things turn out. Anyone who feels an eel should be allowed to remain an eel can no longer afford the luxury of also letting it remain a mystery.

At least we know one thing about the demise of the eel: it's our fault. All the explanations put forward by science to date have something to do with human activity. The closer humanity gets to the eel, and the more it's exposed to the influence of our modern living, the faster it dies. When the International Council for the Exploration of the Sea (ICES) summarized what should be done to save the eel in 2017, it was simultaneously vague and laudably clear: the impact of human activity on the eel should be "as close to zero as possible." We still don't know everything about the threat to the eel, but what we do know is enough to identify the only way of saving it: we have to leave it be.

What we know, for example, is that the eel is struggling with disease, and more so now than before. It's susceptible to, among other things, the herpes virus *anguillae*, a disease first discovered among Japanese eels in captivity, which has since spread through imports to wild eels in Europe. The first Dutch case was identified in 1996; in southern Germany, tests have shown that nearly half of all eels have it.

For some reason, the virus seems to affect only eels—hence its name—and it's an unusually unpleasant disease. The virus can lie dormant in its host for a long time, but once it breaks out, it has a quick and aggressive course. The eel develops bleeding sores around its gills and fins. The cells in the gills die and the blood-filled filaments stick together. Its inner organs become inflamed, rendering the eel tired and lethargic until it can move only slowly and near the surface, until its body finally gives up and it dies.

Eels can also catch the parasite *Anguillicoloides crassus*, a nematode. It, too, was first discovered among Japanese eels and reached Europe in the 1980s, probably piggybacking on live eels imported from Taiwan. In just a few decades, it has since spread across all of Europe and to America. A 2013 study in South Carolina showed that as early as the glass eel stage, 30 percent of eels carried the parasite. The study also indicated the parasite had spread faster due to well-intentioned attempts to save the eel by releasing caught glass eels in new waters.

The nematode is a kind of roundworm that specifically attacks the eel's swimming bladder, causing bleeding, inflammation, and scarification. An infested eel grows more slowly and becomes more susceptible to disease. It moves into shallower waters and can swim only for short distances. The parasite isn't necessarily fatal, but an eel infested with *Anguillicoloides crassus* has very poor prospects of reaching the Sargasso Sea.

What we also know is that the eel is particularly sensitive to pollution. Since it lives for a long time and sits high up the food chain, it's particularly affected by industrial and agricultural toxins. And as with the parasites, the toxins seem to impede the eel's ability to make the journey back to the Sargasso Sea. Eels exposed to PCB, for example, have been shown to develop heart defects and edema and problems storing fat and energy, which makes the long migration virtually impossible. Eels exposed to various pesticides have been shown to be less able to transition from fresh to saltwater. And if appearances are anything to go by, if it's true that fewer silver eels reach their spawning grounds, pollution is at least a likely contributing factor.

Some theories are harder to prove. There are some signs pointing to the eel's falling prey to other predators more often than before, which may not be directly attributable to humans; but it's conceivable that eels that are sick, weakened by toxins and parasites and therefore moving more slowly and closer to the surface, also make easier targets for predators like cormorants, who are plentiful and love feasting on eels.

Some modern threats that researchers consider the most serious, and which are unquestionably caused by humans, are the various physical impediments to the eel's migrations. Locks, sluices, and other artificial means of water regulation can keep young eels from swimming up waterways and mature eels from reaching the sea. And hydroelectric plants, beneficial as they may be for the greater environment, are death to eels. The dams' turbines kill scores of silver eels on their way toward the Atlantic, with some reports claiming that each power plant kills close to 70 percent of all eels trying to pass through. The fish ladders built to circumvent the dams are, by and large, customized for use by the more shallowly inclined salmon.

One old threat to the eel's survival is, of course, fishing, though the severity of its impact has long been the subject of debate. Historically, the eel has been a popular food in many parts of Europe; not only have eel fishermen had their own traditions, tools, and methods, the eel industry has also supported a distinct and in places significant economy. Over the past few decades, exports to Japan—which is now responsible for 70 percent of the world's eel consumption and which, like Europe and America, is feeling the effects of a shrinking eel population—have risen dramatically.

Particularly devastating to the eel's complex life cycle has been the fishing for glass eels. These days, this is primarily done in Spain and France—in the Basque Country, glass eels fried in oil and garlic have become an increasingly expensive delicacy in recent decades—and since they are caught in such large numbers, and at such an early stage of life, the fishing has an outsize impact on the greater population.

A threat that's more difficult to illustrate, but which may nevertheless be the most serious, is climate change. It's an indisputable fact that when the climate changes, both the direction and the strength of the great ocean currents change, which seems to be impeding the eel's migration significantly. Altered currents can make it more difficult for the silver eels to get across the Atlantic and find the right spawning ground. More important, however, is the effect this has on the newly hatched larvae that helplessly drift along the currents to Europe.

When the currents weaken and change course, it likely also affects the location of the spawning grounds within the Sargasso Sea, which means the weightless, transparent larvae may fail to find the current that is supposed to carry them to Europe, or that they are simply carried in the wrong direction. Moreover, climate change can alter the currents' temperature and salinity, which in turn affects the production of plankton on which the larvae feed during their journey.

Several studies point to climate change as a major contributing factor in depressing the number of glass eels reaching the coasts in recent years. It is, if nothing else, an ominous warning signal. It means, after all, that the extremely complicated and sensitive process that is the eel's migration and reproduction, which has functioned for millions of years, has now, in just a few short decades, been fundamentally hobbled.

SO WHAT WILL REMAIN OF THE EEL IF IT GOES EXTINCT? PICTURES, memories, and stories, of course. A riddle that was never fully solved.

Perhaps the eel will become the new dodo. Perhaps it will seem less and less like a real, living creature and more and more like a tragicomic, symbolic reminder of what humankind is capable of in its most oblivious moments.

The dodo was a clumsy, broad-beaked bird that humans first came across at the end of the sixteenth century and had hunted to extinction less than one hundred years later. It was discovered and described for the first time by Dutch sailors on the island in the Indian Ocean that would later be named Mauritius, the only place in the world it ever lived, as far as we know.

It was a large bird, about three feet tall and weighing more than thirty pounds. It had tiny wings, grayish-brown feathers, a bald head with a slightly bent, green-and-black beak. Its legs were yellow and powerful, its rump rounded and wide. It was flightless and moved fairly slowly, but had no natural enemies on the island before humans arrived. Contemporary depictions often ridiculed its appearance, almost caricaturing it; its expressionless eyes like tiny round buttons in its big, bald head, a look of surprise and dim-wittedness on its face.

The earliest mention of the dodo in writing, in a report from a Dutch expedition in 1598, describes it as a bird twice the size of a swan but with the wings of a pigeon. It was also said that it didn't taste particularly good and that its meat was tough no matter how long you cooked it, but that the belly and breast were at least edible.

Which is of course what the Dutch sailors did to the dodo: they ate it. It was very easy to catch, after all. It's said the birds didn't even try to escape when the sailors approached them. They were fat and rich in meat; three or four of them was enough to feed a whole crew. Dodoes were described as non-chalant and unperturbed, as though utterly unable to imagine that another creature could potentially constitute a threat. A drawing from 1648 shows sailors merrily beating the clumsy birds to death with big sticks. Their fate was not only to be the dinner of hungry Dutch sailors, however; humans also brought other invasive species to the island: dogs, pigs, and rats that competed for space and food and raided the dodoes' nests, eating their eggs and chicks.

In the summer of 1681, Benjamin Harry, a sailor, men-tioned in his diary that he had seen a dodo in Mauritius. That represents the last documented sighting of a living specimen. The dodo he saw was, if the story is to be believed, the very last one. Then it was dead, extinct, and all that remained were fading memories.

For a while, the dodo was forgotten or depicted as a vaguely mythological creature, rather than a real animal. Some doubted it had ever existed at all. When Alexander Melville and Hugh Strickland published their book *The Dodo and Its Kindred*, the most exhaustive description of the dodo at the time, in 1848, they were forced to admit that information about this bird, which had been extinct for more than 160 years, was scarce, to say the least. "We possess only the rude descriptions of unscientific voyagers, three or four oil paintings, and a few scattered osseous fragments, which have survived the neglect of two hundred years. The paleontologist has, in many cases, far better data for determining the zoological characters of a species perished myriads of years ago, than those presented by a group of birds, several species of which were living in the realm of Charles the First."

They were at least able to establish that the closest living relative of the dodo is the pigeon; modern DNA testing has since confirmed their findings. Other than that, though, Melville and Strickland didn't contribute much to our overall understanding of the dodo. That this idiosyncratic creature lived where it lived and only there was not at all strange, they argued. The temporal and geographical distribution of species had nothing to do with environment or climate, and certainly not with evolution. It was the "Creator's" way of preserving "the ever vacillating balance of Nature." That the dodo had become extinct was, consequently, not surprising. "Death," they wrote, "is a Law of Nature in the Species as well as in the Individual."

In time, however, we would learn a lot more about the dodo. In 1865, the first fossil was found, and science began to take a greater interest in its unique fate, both as the odd bird it had been and as an example of humankind's boundless and irrevocable impact on all life on this planet. Since the end of the nineteenth century, countless books have been written about the dodo. Lewis Carroll's *Alice's Adventures in Wonderland* has made it iconic; it is doubtless one of the most widely recognized extinct species today. Furthermore, the dodo has become a symbolic creature, not only as a cautionary example of the reckless cynicism of humankind, but also as a metaphor for something outdated and obsolete. A dodo is a person who is stupid and clumsy and incapable of adapting to a new era, someone who has been rejected and forgotten, become irrelevant.

"Dead as a dodo," as the expression goes. It may be that we will eventually say "dead as an eel" instead.

THAT MAY BE PREFERABLE TO OTHER CONCEIVABLE FATES. PERHAPS the eel will instead become something like Steller's sea cow, a quickly fading memory of something odd and unfamiliar.

Steller's sea cow was the name of a marine sirenian first described in the middle of the eighteenth century by the German scientist Georg Wilhelm Steller. It was a gigantic mammal, a languid, slow herbivore like its closest relatives, the dugong and the manatee. It had thick, bark-like skin and an undersize head relative to its enormous body, two small arms in front, and a whalelike tail in the back.

Georg Wilhelm Steller first spotted the animal during an expedition led by the Danish-Russian explorer Vitus Bering, in what would eventually be named the Bering Sea. It was Bering's second expedition to the mostly unexplored region, and his mission, given to him by the Russian navy, was to sail across the sea and map the west coast of North America. Steller had on his own initiative, driven by curiosity and a thirst for adventure, traveled east through Russia to join Bering. He'd studied theology, botany, and medicine at the University of Wittenberg, accompanied a caravan of wounded Russian soldiers to Saint Petersburg, and secured a position as the personal physician to the archbishop of Novgorod. He was almost thirty and just married when he set off through vast Siberia in the winter of 1737, with his sights set on the Kamchatka Peninsula, where Vitus Bering was preparing for his expedition.

On May 29, 1741, the ship *Saint Peter* set off from Okhotsk with a crew of seventy-seven. It would be a disastrous journey in most respects. Almost immediately, the expedition encountered difficult weather, lost contact with its sister ship, the *Saint Paul*, and was forced to veer south across the sound toward the North American coast. Once they reached Alaska, the crew was already in poor shape, and many were suffering from scurvy. On top of everything else, Bering and Steller didn't get along. Bering wanted to hurry up and map as much of the coast as they could and then turn back before the arrival of the autumn storms. Steller, for his part, wanted to do what he had come there to do: study the flora and fauna.

After about two months at sea, Bering developed scurvy, and it was decided the ship would immediately turn around and return to Kamchatka. But a violent storm intercepted them, and the ship ran aground on the reefs off an island that no one knew existed. There, in the breakers off the strange land, while most of the crew were lying unconscious in the damaged ship and the corpses of the already perished were being thrown overboard, an eager Steller immediately started planning his excursions. He had animals and plants to study. And it was there, on the island that would later be named Bering Island, just east of Kamchatka, that Georg Wilhelm Steller on November 8, 1741, first spotted a large herd of the previously unknown species of sea cow resting at the water's edge.

It was clearly a magnificent sight, and Steller described the animals that would later be named for him in detail. From the navel up, they looked like large seals, he wrote, but below the navel they were more akin to fish. Their heads were round and not at all dissimilar from the buffalo's. Their eyes were, despite the size of the animal, no larger than a sheep's and had no eyelids. Their ears were hidden in the folds and furrows of their thick skin. Other than the wide tail, it lacked fins, which set it apart from the whale. "These animals live like cattle in herds in the sea," Steller wrote. "They do nothing but eat."

Steller not only described what the exotic sea cows looked like, what they ate, how they behaved, and how they reproduced. He also described in equal detail how fat and tasty they were, and that they were so plentiful they could have fed all of Kamchatka. He wrote that they showed no fear of humans at all. They didn't try to escape when approached, and their only response when the starving members of the expedition caught them with large iron hooks and cut meat out of them while they were still alive was to sigh quietly.

What the sea cows lacked in survival instinct, Steller declared, they made up for in touching displays of empathy.

> *Signs of a wonderful intelligence . . . I could not*
> *observe, but indeed an uncommon love for one another,*
> *which even extended so far that, when one of them*
> *was hooked, all the others were intent upon saving*
> *him. Some tried to prevent the wounded comrade from*
> *[being drawn on] the beach by [forming] a closed*
> *circle [around him]; some attempted to upset the yawl;*

others laid themselves over the rope or tried to pull the harpoon out of [his] body.

One of the males, Steller wrote, even returned two days in a row to check on one of the females who lay dead on the beach. "Nevertheless, no matter how many of them were wounded or killed, they always remained in one place."

The encounter with the languid but loving sea cows was not just a profound experience for Georg Wilhelm Steller; it was a biological sensation. Sirenians, mammals that are in fact more closely related to the elephant than the seal or the whale, are normally found only in tropical waters. This species lived on a cold, barren island far in the unexplored northern part of the Pacific Ocean, and apparently only there. Steller's sea cow was yet another powerful example of the complexity of evolution and the mesmerizing diversity of this world. A strange living wonder in one of the world's most inhospitable places.

But like sirens, Steller's sea cow brought destruction on both its discoverers and itself. Vitus Bering died on the island on December 8 and was buried in the sand by the water's edge. About half the crew shared his fate. Steller himself made it. He and the other survivors wintered on Bering Island, surviving by catching sea otters, whose flesh they ate raw. In the spring, they managed to build a new ship from the wreckage of the *Saint Peter*, and in August 1742, more than a year after they set off, they returned to Kamchatka, emaciated and decimated. Georg Wilhelm Steller published his observations, and was able to tell the world about the strange northern sirenians, but soon after lost himself to drink and died, just thirty-seven years old, in Tyumen', Russia, in 1746.

And Steller's sea cows perished, too. Russian hunters followed in Bering's footsteps and found the languid animals to be easy prey. In 1768, only twenty-seven years after being discovered by Steller, the last sirenian was killed in the Bering Sea, and today few people even know it ever existed. It vanished from humankind's awareness and realm of knowledge with a quiet sigh, docilely accepting its fate. Unlike the dodo, it didn't even pass into the vernacular.

BUT THE EEL IS NEITHER A DODO NOR A SEA COW. FIRSTLY, IT'S NOT isolated on some island in the Indian Ocean or the Bering Sea. Secondly, it has survived humanity for too long to come to that kind of abrupt end. And surely all the energy spent on understanding it over the centuries can't have been for naught?

Because there are, after all, a lot of people doing their best to help the eel. Just as the life cycle of the eel has for centuries aroused the curiosity of science, many scientists working today consider its demise the most important challenge they currently face.

Some of the alarms sounded by researchers and organizations like ICES and IUCN have been taken very seriously. At least in Europe. In 2007, the European Union adopted a management plan containing a series of radical proposals to try to save the eel. Every member country committed to implement measures to ensure that at least 40 percent of all silver eels can reach the sea, by, for example, limiting fishing and building alternative passes to circumvent dams and power stations. All exports to non-European countries, such as the insatiable Japanese market, have been banned (though illegal exports are still assumed to be substantial), and anyone fishing for glass eels must set aside at least 35 percent of the catch for reintroduction into the wild. In the same year, 2007, Sweden's National Board of Fisheries banned any form of eel fishing in Sweden, with the exception of professional eel fishermen with special permits, or in fresh water upstream from the third migration barrier.

At first, the measures seemed to be having an effect. In the years that followed, the European eel did seem to recover slightly. There was, above all, an increase in the number of glass eels arriving from the Sargasso Sea, and for the first time in a long time, the people who care about eels could allow themselves a quantum of optimism.

But since 2012, the trend has reversed and the rate of recovery has leveled off. The slight uptick seems to have been a temporary exception, and the goals set up in the EU's management plan have remained far from achieved. On the whole, the eel's situation is at least as dire today as it was before 2007.

We seem to be stuck in a "utopian deadlock," as Willem Dekker, an eel expert at the Swedish University of Agricultural Sciences in Uppsala, wrote in a summary of the situation in 2016. The hopefulness we had been feeling for some time turned out to rest on unrealistic expectations. The measures put in place to save the eel, Dekker claimed, are not only insufficient, they also risk becoming a placating form of misdirection. As long as we cling to what we think we know, what we believe to be right, the eel's situation will never improve, but instead worsen.

And while the problem continues to be debated, time passes.

In the autumn of 2017, the EU's agriculture and fisheries ministers were due to set new fishing quotas, and the European Commission's surprisingly radical proposal was to ban all eel fishing in the Baltic Sea. Sweden supported a blanket ban at first, but when no other country joined the cause, it chose to abandon it. It's important to be open to negotiation, the Swedish minister for rural affairs, Sven-Erik Bucht, stressed; he, like so many others, apparently had fonder feelings for fish other than the eel. If we choose to stand up for the eel, we give up our chance to protect other species, he argued. "No one is going to be able to take the salmon's side." Once the decision had been made, there were, consequently, reductions in the quotas for salmon, cod, herring, and plaice, while the eel could continue to be fished much as before.

It took another year, until December 2018, before the EU decided to implement a union-wide ban on eel fishing, including in the Mediterranean and along the Atlantic coast. But the ban covers only three months of the year, and the glass eel is not yet included in it.

And thus, the eel populations continue to decline, while decisions about what to do to help it are punted down the road. Until we know more. Or until there's nothing left to know.

IS IT POSSIBLE TO IMAGINE A WORLD WITHOUT EELS? IS IT POSSIBLE to erase a creature that has existed for at least forty million years, that has survived ice ages and seen continents drift apart, that when humans found their place on this planet had already been waiting for us for millions of years, that has been the subject of so many traditions and celebrations and myths and stories?

No, is the instinctive answer, that's not how the world works. What exists, exists, and what doesn't exist is always in some ways unimaginable. Imagining a world without eels would be like imagining a world without mountains or oceans, air or soil, bats or willow trees.

Yet at the same time, all life is changeable, and we will all change one day, and it was probably, at some point, at least for a few people, just as difficult to imagine a world without the dodo or without Steller's sea cow. Just as I couldn't, once, imagine a world without Nana or Dad.

And yet they're both gone now. And the world is still here.

18

In the Sargasso Sea

I don't remember the last time we went eel fishing, but as time passed, it happened less and less frequently. Not because the eel lost any of its mystery, but perhaps because other mysteries became more important. Our closed little world down by the river found it increasingly difficult to compete with all the other worlds that gradually opened up. This was, of course, a predictable development. People grow up, change, leave, transform, stop fishing for eel. With all the symbolic metamorphoses we go through, some things are inevitably lost.

As a teenager, I sometimes took friends down to the stream. Dad stayed home. We brought beer and an air gun, and when we caught an eel, we tried to shoot it in the head. We took turns, shooting and missing and shooting again. I brought the eels home to Dad, who was furious when he almost broke his teeth on diabolo pellets. I think he felt we were being disrespectful, to him but maybe even more so to the eel.

Dad went down to fish by himself sometimes, but not as often. I finished school and started working. I went out on the weekends. We grew apart, not because of conflict or rejection but simply because everything changed of its own accord. The current that had once swept Dad with it to a new place now seemed to be carrying me away from him. When I was twenty, I moved away and ended up at what the current seemed to consider my final destination: university.

If the eel was our shared experience, university was the opposite, a manifestation of all the things we didn't share. A strange place, very different from everything I was used to. A place where memories manifested as large buildings and people spoke in abstractions, in a language I didn't understand, where no one seemed to work and everyone was busy self-actualizing. And I was fascinated by it, if slightly reluctantly. I let myself absorb the environment and the culture and learned how to mimic all the exotic social codes. I carried my books around as though they were identity papers and I learned to answer succinctly and defensively whenever anyone asked where I was from. I suppose I figured the smell of asphalt would expose me as a stranger in the academic corridors.

But at some point every summer, I would go back home and we'd head down to the stream to fish for eel. We'd abandoned the spillers and the trap by then and instead switched to a more modern form of bottom fishing. We used regular hazel rods with tackle consisting of a large single hook and a heavy sinker. We baited our hooks with worms and let them sink to the streambed. Dad had made rod holders out of heavy metal pipes, which we pushed into the ground so the rods stood erect like masts against the night sky. We brought foldable camping chairs and put bells at the tip of the rods that would jingle when we had a bite. Then we sat there, well into the night, listening to the monotonous sound of the rapids, watching the shadow of the willow tree lengthen and the bats swerve nimbly around our rods as they flitted past. We drank coffee and talked about eels we'd caught and eels we'd lost and not much else. Despite everything, I never grew tired of it.

Eventually, my parents bought a cabin. It was a red wooden cabin, small and not particularly pretty, with no indoor plumbing and a well full of dirty water. But it was built next to a small lake, surrounded by forest on every side, with big stands of reeds in which mute swans and great crested grebes nested. Almost every day, herons and ospreys would fly over the lake, and in the evenings, the sun would set like a big ball of fire behind the spruce trees on the other side. Mum and Dad loved that place and spent as much time there as they could.

There was a small plastic boat that belonged to the cabin, and on my visits, we would fish on the lake. Mostly pike and perch. We rowed around, exploring the lake, which was larger than it initially appeared. The cabin was located on the east side, and at the southern end was a large, shallow patch of reeds, where you could hear pike splashing about at dusk. A small stream emptied into the lake at its northern end; the perch hunted there around the clock. In the west, the lake stretched into a long, narrow arm chock-full of reeds, water lilies, and small, grassy islets. We figured that was where the biggest pike lived.

One night, we were sitting in the cabin, gazing out across the water. The lake had flooded and climbed several yards up the lawn, and suddenly big, powerful tail fins broke the surface, right at the edge of the grass. They swayed this way and that like dark pennants in the moonlight. They were tench, we decided eventually, and we fished for them the way we used to fish for eel: ledgering with hazel rods with bells on their tips. I caught one that weighed almost three and half pounds; it was dark and slimy and had tiny, almost invisible scales. We caught bream, too, sluggish, clumsy fish that somewhat resignedly let themselves be pulled out of the water.

But we never caught a single eel, which as time wore on seemed more and more mysterious.

"There must be eels here," Dad would say. All the signs indicated as much. The lake was shallow and the lake bed muddy; there was plenty of vegetation and rocks to hide among, and the water was teeming with small fish. The stream that emptied into the lake would present no challenge to a migrating eel, and it was connected to the stream we had always fished for eel in, which was only about twenty miles away.

"I don't understand why we never catch any," Dad would say. "There just have to be eels here."

And yet we never so much as glimpsed a single one. As if to remind us of what it had once meant to us, it hid in the shadows. Eventually, we started wondering if it existed at all.

DAD FELL ILL IN EARLY SUMMER THE YEAR HE TURNED FIFTY-SIX. THAT something was wrong had been known for a long time. He'd been in pain and had eventually gone to see a doctor, who had in turn referred him to the hospital. They had done X-rays and tests and eventually determined what the problem was: a

IN THE SARGASSO SEA

large, aggressive tumor. Why dad was sick was explained by a doctor who told us about the clear correlation between working with asphalt and the kind of cancer he had. The warm steam from the asphalt had eventually penetrated to the depth of his very core, and there was now literally no way of ever getting it out.

He had surgery as summer turned to autumn; it was a big, complicated surgery, and we were well into winter before he could leave the hospital. For months he lay in bed attached to an IV, unable to eat or even enjoy his snus, and we would come to visit and watch in silence when the staff made him get out of bed and walk up and down the hallway, leaning over a walker. He was pale and thin under his hospital gown. It was the first time I'd seen him really weak.

It was also there, one day in the hospital cafeteria, while Dad was in his room, drowsy from the morphine he'd been given, that my mom told me what I should have understood much earlier. My grandfather, the person I had always called Grandad, wasn't my father's father. His biological father was someone else entirely, someone none of us knew, not even Dad. My grandmother had met that man when she was about twenty. She had become pregnant and had a child, and the man had wanted nothing to do with her or his son. That was all we knew about him, aside from his first name, which was also my father's middle name.

Why hadn't I realized sooner? How could I have missed that? I knew Dad had spent his first years living with Nana's parents. I knew he'd been looked after by Nana's sisters when she was at work at the rubber factory in town. I'd heard about when my great-grandmother died, when Dad was just a couple of years old, and when they moved from the contract worker's cottage to their own house. For some reason, I just hadn't put two and two together.

Nana hadn't met the person I would eventually call Grandad until my dad was about seven. They'd been an item for only a short while when Dad had come home inconsolable after his first day of school. All the children in his new class had been asked to tell the others who their fathers were. But Dad didn't know. He hadn't been able to say anything, and maybe he'd realized for the first time that our origin is something that affects us, whether we want it to or not, and that a person who doesn't know his origin will always be a little bit lost. If you don't know where you came from, you can't know where you're going. The journey away from home and back to it follow the same given route.

Soon after that first day of school, my grandparents got engaged. They were married just a few weeks later, quickly and without fuss, with Nana's sisters as the only witnesses.

Grandad, the person I would continue to call Grandad, had from the outset treated Dad like his own son, and it seems Dad made a decision then and there. His origin was a riddle he would choose the answer to. He'd spent his first seven years without a father, and now he suddenly had one. The invisible figure who had passively occupied that role until then didn't interest him in the slightest, and the reason he'd never told us the truth was that he didn't want us to feel any doubt about the way things were. Our grandad was the kind, decent man who, unlike the invisible man, had actually been there. At some point, Dad had simply decided that his, and consequently our, origin was there with him, on the farm by the stream, and that was the truth, in every way that mattered. Not even now, when he was sick and nothing was certain, did Dad talk about it, and we never asked him.

The surgery, and the almost six months of bed rest, gave Dad four more years. Four years of slow recovery before the tumors would come back, each time more brutal. First a relapse and another autumn of surgeries, complications, pain, and several months in the hospital. Then a second relapse; by then, he was so weakened there was no point in fighting.

Dad had turned sixty by then. I was sitting with him at the house, watching TV, one early evening. He was relaxing in the black armchair and had put his feet up on a stool in front of him; he was tired but in a good mood. We didn't know then that the tumor had come back; we didn't know anything about what was once again lurking inside his body. At least I didn't.

"Are the water levels still high out by the cabin?" he asked.

"No, the water's subsiding, it only just about covers the jetty now."

"But the jetty's still there, right? It didn't move?"

"No, it looks fine, we did a good job securing it. It's going to take something big to shift it now."

"Sure, but how many times have we said that?"

He turned his head and looked at me. "So, have you been doing any fishing?" he asked, and that's when I realized his eyes looked different. The whites had gone yellow, had acquired a grayish-yellow tone, like an old sheet of paper that had turned dirty and matte; the yellow surrounded his black pupils like a thick fog. I looked him in the eyes for a split second, and I must have reacted somehow, because he looked away and turned back to the TV; I sat next to him in silence, staring straight ahead, without really knowing what had just happened.

We talked some more, but each time I looked at him it was as if he were trying to avoid my gaze. He turned his head away, as though hiding something from me, and I remembered a time when I was little and we were sitting around the kitchen table. It was in the middle of winter and snowy and cold outside; Dad was wearing a yellow knit hat with a blue crown on it, and when he took it off the skin of his forehead was the same shade of yellow as the hat. "I've got jaundice," he said and chuckled, but I didn't understand it was a joke. I asked Mum what jaundice was and she said it was a disease of the liver and that it could be fatal and I went scared and quiet. I thought Dad was dying, and I had no words to express my fear. When he laughed and explained he was kidding and it was just the hat rubbing off on him, I didn't dare to believe it. I had realized that if other people could fall ill and even die, then why not my dad? Why not me?

As we watched the television, darkness fell outside and Dad grew tired, but I could feel him fighting it. He wanted to stay up a while longer. He didn't want to acknowledge the fatigue that had taken over his body, or admit something was wrong. So he sat there, listening and talking, in a low, soft voice, and suddenly, almost in the middle of a sentence, he closed his eyes and fell asleep. He sat there in his reclining chair, completely still with his eyes closed, breathing deeply and heavily, as though he'd just clocked out. I sat alone in the chair next to him; eventually, I turned back to the TV and waited, without really knowing what I was waiting for.

A short while later—ten seconds, twenty seconds—he opened his eyes again, looked at me, and tried to smile. "I must've drifted off," he said.

A few weeks later, I visited him in the hospital; it was two days after midsummer, and nothing was hidden anymore. It's back, the doctor had explained; this time, the tumor was attacking the liver. When we asked what could be done, the young, serious doctor spread his hands and shook his head.

I think Dad understood it better than I did. "I'm not going to make it this time," he said; I tried to argue but couldn't find the words. "I hope you'll want to keep the cabin," he said—at least I could promise him that. A few days later, he was transferred to hospice and sank into unconsciousness.

THE THIRD OF JULY WAS A THURSDAY. THE WEATHER WAS WARM AND stifling. We were sitting in Dad's small room at the hospice with the patio door open onto a plot of grass. Beyond the lawn, behind some trees, there was a small pond, where a heron stood, turning its head this way and that, peering out across the still surface.

It had been a difficult night. Dad had made a lot of noise, whimpering and groaning as though he were worried and in pain, even in his unconscious state. Mom, who spent her nights on a cot in his room, had barely slept a wink.

That morning, when I arrived, he was calmer; I sat alone by his bedside, holding his hand. It was warm and damp; his rough fingers were stiff like bits of wood. He was quiet and completely still. I listened to his breathing, faint and irregular; between each breath, the seconds stretched out like eternities.

And I wondered, for the first time, how you recognize death. How do you know when it has come?

"When the heart stops beating," is probably what most people would say. When the last breath leaves the body and everything is still. That's how we've traditionally thought about the moment of death; heartbeats and breathing are necessary to live, and thus we have a clear boundary between life and death. The exact second the heart stops beating is the moment death occurs. The time of death can be definitively established. Like a candle being blown out.

But that's not necessarily what death looks like. Hearts don't usually stop beating from one second to the next; instead, they gradually beat slower and more irregularly. They can stop beating and then start again. Blood pressures drop, oxygen levels fall. Rather than suddenly replacing life, death seeps slowly into it.

In Sweden, legal death has nothing to do with heartbeats and breathing. According to Swedish law, a person is alive as long as his or her brain shows some form of activity. The first paragraph of the law outlining the criteria to determine death in a human states that "a person is considered dead when there is complete and irrevocable cessation of all brain function."

It's worded that way partly to make it easier to harvest organs for transplant from a brain-dead person on a respirator, but it's also a definition that puts a kind of value on life. Because it means life isn't simply a biological function but rather something linked to consciousness—if not to waking consciousness, then at least to the theoretical ability to perceive things, to feel or dream.

That ability doesn't seem to be entirely dependent on heart-beats or breathing. In 2016, a research team from the University of Western Ontario in Canada studied the moment of death in four patients. After all life support had been disconnected, brain activity was measured with electrodes. In three of the four patients, all brain activity had ceased before the heart stopped beating, in one of them no less than ten minutes before. But in the fourth patient, the opposite was true. The instruments showed brain activity ten full minutes after the last heartbeat. What was going on in there? What did those crackling peaks on the EEG curve consist of? Images? Feelings? Dreams?

In another study, conducted by Lakhmir Chawla, an American intensive care physician, heightened brain activity was recorded at the moment of death. Chawla noted increased activity for thirty seconds to three minutes from the moment the heart stopped beating in seven patients. The patients, who had been in a state of deep unconsciousness, had, in the final moments of life, suddenly demonstrated levels of brain activity almost equal to those of a fully conscious person. Since he published his report in 2009, Lakhmir Chawla has observed the same phenomenon in more than a hundred dying patients, and though his results have been questioned, they seem to lend some support to the notion of what are commonly referred to as near-death experiences. Perhaps there are mental states we don't know about and which we will never fully understand until someone can tell us about them from beyond the grave. And perhaps these mental states are completely dissociated from the things we usually use to quantify life—heartbeats and breathing, but also time itself. At least that is a theory put forward by Arvid Carlsson, who received the Nobel Prize in medicine in 2000. Perhaps, he commented in an article, we experience at the moment of death a state that is completely dissociated from time.

"And what is that?" he asked. "It's eternity. Right?"

My dad had no electrodes connected to his head. I didn't know if there was any level of awareness left in him that warm morning, or what he might have been feeling or dreaming about if there was. Nor did I know how long I'd been sitting there—I had eventually lost all sense of time—but when I squeezed his hand harder, I suddenly realized I hadn't heard him breathe in a while. I called a nurse, who came in quickly and reached for his wrist to feel his pulse. I watched her, still holding his other hand in mine. She looked back at me and nodded quietly.

THE NEXT DAY, WE WERE SITTING OUTSIDE THE HOUSE, LISTENING TO the church bells ringing for Dad less than half a mile away. We were sitting on the lawn next to the apple tree, in front of the greenhouse where the tomatoes were starting to turn red, in the exact spot where we had planted the pitchfork to drive the worms out of the ground, where we had painted the rowboat, and where Dad had put out the eel trap one day. The bell tolled dully and ponderously from what sounded like endlessly far away.

A week or so later, after the funeral, we went out to the cabin. It was another warm, stifling summer's day. The grass was dry and in need of mowing. The osprey soared above the lake, which lay completely still in the blazing sunlight. I stood by the water's edge with a fishing rod in my hand, staring at the bobber. Someone called me; I put the rod down on the grass, the bobber still in the water. When I came back a few minutes later, I realized something was about to pull the entire rod into the lake. It was sliding quickly through the grass, the line taut; I grabbed it at the last second and immediately felt the undulating resistance of a fish. I had time to think the feeling was familiar before the fish set off toward the water lilies. Then it suddenly turned and swam back toward the shore, and before I could react, the line had disappeared in among the big rocks next to the shoreline. And there it got stuck.

For a moment, time stood still. The taut line and the tiny, struggling movements. I coaxed and pulled, and the rod bent like a reed; I took a few steps to the side to find a new angle, tugging so hard on the line the nylon sang. I thought there were only two ways out of this situation and both had its losers, and I cursed under my breath and finally sank onto my knees, clutching the line, peering down into the murky water.

I know it was an eel because I saw it. It slowly slithered up out of the shadows and came toward me. It was large and a pale shade of gray, with black button eyes, and it looked at me as if to make sure I could see it. I let go of the line and saw the hook come out just as the eel reached the surface, then it turned and slid back into the hidden depths.

For a while, I just sat there by the water's edge. Everything was quiet and the lake completely still; the sun sent a white sheen spreading across the water and everything beneath the surface was hidden, as though behind a mirror. What lay hidden underneath was a secret, but now it was my secret.

Sources

The excerpt from a poem by Seamus Heaney on page v is from "A Lough Neagh Sequence," from the collection *Door into the Dark* (New York: Oxford University Press, 1969).

3: ARISTOTLE AND THE EEL BORN OF MUD

Aristotle. *Historia animalium* (*The History of Animals*). Translated by D'Arcy Wentworth Thompson. Oxford: Clarendon Press, 1910.

Homer. *The Iliad*. Translated by Robert Fitzgerald. Garden City, NY: Anchor, 1974.

Lennox, James. "Aristotle's Biology." In *Stanford Encyclopedia of Philosophy*. Stanford University, Metaphysics Research Lab, Center for the Study of Language and Information. Revised January 31, 2016. https://plato.stanford.edu/entries/aristotle-biology/.

Marsh, M. C. "Eels and the Eel Question." *Popular Science Monthly* 61 (September 1902).

Prosek, James. *Eels: An Exploration, from New Zealand to the Sargasso, of the World's Most Mysterious Fish*. New York: Harper, 2010.

Schweid, Richard. *Consider the Eel: A Natural and Gastronomic History*. Chapel Hill: University of North Carolina Press, 2002.

Walton, Izaak. *The Compleat Angler*. London: 1653.

5: SIGMUND FREUD AND THE EELS OF TRIESTE

Cairncross, David. *The Origin of the Silver Eel—With Remarks on Bait & Fly Fishing*. London: G. Shiel, 1862.

Eigenmann, Carl H. "The Annual Address of the President—The Solution of the Eel Question." *Transactions of the American Microscopical Society* 23 (May 1902).

Freud, Sigmund. *The Letters of Sigmund Freud to Eduard Silberstein, 1871–1881.* Edited by Walter Boehlich. Cambridge, MA: Belknap Press, 1990.

Marsh, M. C. "Eels and the Eel Question." *Popular Science Monthly* 61 (September 1902).

Simmons, Laurence. *Freud's Italian Journey.* Amsterdam: Rodopi, 2006.

Whitebook, Joel. *Freud: An Intellectual Biography.* New York: Cambridge University Press, 2017.

7: THE DANE WHO FOUND THE EEL'S BREEDING GROUND

Eigenmann, Carl H. "The Annual Address of the President—The Solution of the Eel Question." *Transactions of the American Microscopical Society* 23 (May 1902).

Garstang, Walter. *Larval Forms and Other Zoological Verses.* 1951.

Grassi, Giovanni Battista. "The Reproduction and Metamorphosis of the Common Eel (*Anguilla vulgaris*)." *Proceedings of the Royal Society of London*, January 1896.

Marsh, M. C. "Eels and the Eel Question." *Popular Science Monthly* 61 (September 1902).

Poulsen, Bo. *Global Marine Science and Carlsberg: The Golden Connections of Johannes Schmidt (1877–1933).* Boston: Brill, 2016.

Schmidt, Johannes. "The Breeding Place of the Eel." *Philosophical Transactions of the Royal Society of London* 211 (1923), 179–208.

Tsukamoto, Katsumi, and Mari Kuroki, eds. *Eels and Humans.* New York: Springer, 2014.

9: THE PEOPLE WHO FISH FOR EEL

www.alarv.se.

www.alakademin.se.

Prosek, James. *Eels: An Exploration, from New Zealand to the Sargasso, of the World's Most Mysterious Fish.* New York: Harper, 2010.

Schweid, Richard. *Consider the Eel: A Natural and Gastronomic History.* Chapel Hill: University of North Carolina Press, 2002.

Tsukamoto, Katsumi, and Mari Kuroki, eds. *Eels and Humans.* New York: Springer, 2014.

11: THE UNCANNY EEL

The Bible, Revised Standard Version.

Eco, Umberto, ed. *On Ugliness.* Translated by Alastair McEwen. New York: Rizzoli, 2007.

Freud, Sigmund. *Das Unheimliche.* 1919.

Friedman, David M. *A Mind of its Own: A Cultural History of the Penis.* New York: Free Press, 2001.

Grass, Günter. *The Tin Drum.* Translated by Ralph Manheim. New York: Pantheon, 1961.

Hoffmann, E. T. A. "The Sandman." 1816.

Jentsch, Ernst. *Zur Psychologie des Unheimlichen.* Psychiatrisch-Neurologische Wochenschrift: 1906.

Myśliwiec, Karol. *The Twilight of Ancient Egypt: First Millennium B.C.E.* Translated by David Lorton. Ithaca, NY: Cornell University Press, 2000.

Nilsson Piraten, Fritiof. *Bombi Bitt och jag.* Stockholm: A. Bonnier, 1932.

Swift, Graham. *Waterland.* New York: Poseidon Press, 1983.

Vian, Boris. *The Foam of Days.* 1947.

Winslow, Edward, and William Bradford, *Mourt's Relation: A Journal of the Pilgrims at Plymouth.* London: John Bellamie, 1622.

13: UNDER THE SEA

Carson, Rachel. *The Sea around Us.* New York: Oxford University Press, 1951.

———. *Silent Spring.* Boston: Houghton Mifflin, 1962.

———. *Under the Sea-Wind.* New York: Simon & Schuster, 1941.

Jabr, Ferris. "The Person in the Ape." *Lapham's Quarterly* 11, no. 1 (Winter 2018).

Lear, Linda. *Rachel Carson: Witness for Nature.* New York: Henry Holt, 1997.

Nagel, Thomas. "What Is It Like to Be a Bat?" *Philosophical Review* 83, no. 4 (October 1974): 435–50.

15: THE LONG JOURNEY HOME

Carson, Rachel. *Under the Sea-Wind.* New York: Simon & Schuster, 1941.

Inoue, Jun G., Masaki Miya, Michael Miller, et al. "Deep-Ocean Origin of the Freshwater Eels." *Biology Letters* 6, no. 3 (June 2010): 363–66.

Munk, Peter, Michael M. Hansen, Gregory E. Maes, et al. "Oceanic Fronts in the Sargasso Sea Control the Early Life and Drift of Atlantic Eels." *Proceedings of the Royal Society B* 277 (June 2010): 3593–99.

Prosek, James. *Eels: An Exploration, from New Zealand to the Sargasso, of the World's Most Mysterious Fish.* New York: Harper, 2010.

Righton, David, Håkan Westerberg, Eric Feunteun, et al. "Empirical Observations of the Spawning Migration of European Eels: The Long and Dangerous Road to the Sargasso Sea." *Science Advances* 2, no. 10 (October 2016): https://doi.org/10.1126/sciadv.1501694.

Schmidt, Johannes. "The Breeding Place of the Eel." *Philosophical Transactions of the Royal Society of London B* 211 (1923): 179–208.

Swift, Graham. *Waterland*. New York: Poseidon Press, 1983.

Tesch, Friedrich-Wilhelm. *Der Aal: Biologie und Fischerei*. Hamburg: P. Parey, 1973.

———. "The Sargasso Sea Eel Expedition 1979." *Helgoländer Meeresuntersuchungen* 35, no. 3 (September 1982): 263–77.

16: BECOMING A FOOL

The Bible, Revised Standard Version.

Jerkert, Jesper. "Slagrutan i folktro och forskning." *Vetenskap eller villfarelse*. Edited by Jesper Jerkert and Sven Ove Hansson. Leopard förlag: 2005.

17: THE EEL ON THE BRINK OF EXTINCTION

Carson, Rachel. *Silent Spring*. Boston: Houghton Mifflin, 1962.

Castonguay, Martin, Peter V. Hodson, Christopher Moriarty, et al. "Is There a Role of Ocean Environment in American and European Eel Decline?" *Fisheries Oceanography* 3, no. 3 (September 1994): 197–204, https://doi.org/10.1111/j.1365-2419.1994.tb00097.x.

Castonguay, Martin, and Caroline M. F. Durif. "Understanding the Decline in Anguillid Eels." *ICES Journal of Marine Science* 73, no. 1 (January 2016): 1–4, https://doi.org/10.1093/icesjms/fsv256.

Gärdenfors, Ulf. *IUCN:s manual för rödlistning samt riktlinjer för dess tillämpning för rödlistade arter i Sverige*, 2005.

Hume, Julian P. "The History of the Dodo *Raphus cucullatus* and the Penguin of Mauritius." *Historical Biology* 18, no. 2 (2006): 69–93.

Jacoby, D. and M. Gollock, "On the European Eel." www.iucnredlist.org.

Kolbert, Elizabeth. *The Sixth Extinction: An Unnatural History*. New York: Henry Holt, 2014.

Lear, Linda. *Rachel Carson: Witness for Nature*. New York: Henry Holt, 1997.

Melville, Alexander, and Hugh Strickland. *The Dodo and Its Kindred; or, The History, Affinities, and Osteology of the Dodo, Solitaire, and Other Extinct Birds of the Islands Mauritius, Rodriguez, and Bourbon*. London: Reeve, Benham, and Reeve, 1848.

Steller, Georg Wilhelm. "Steller's Journal of the Sea Voyage from Kamchatka to America and Return on the Second Expedition, 1741–1742." *American Geographical Society Research Series* 2 (1925).

Tremblay, V., C. Cossette, J. D. Dutil, G. Verreault, and P. Dumont. "Assessment of Upstream and Downstream Pass Ability for Eels at Dams." *ICES Journal of Marine Science* 73, no. 1 (January 2016): 22–32, https://doi.org/10.1093/icesjms/fsv106.

Wake, David, and Vance Vredenburg. "Are We in the Midst of the Sixth Mass Extinction? A View from the World of Amphibians." *Proceedings of the National Academy of Sciences* 105 (August 2008): 11, 466–73.

18: IN THE SARGASSO SEA

Norton, L., R. M. Gibson, T. Gofton, et al. "Electroencephalographic Recordings During Withdrawal of Life-Sustaining Therapy until 30 Minutes after Declaration of Death." *Canadian Journal of Neurological Sciences* 44, no. 2 (March 2017): 139–45, https://doi.org/10.1017/cjn.2016.309.

Snaprud, Per. "Dödsögonblicket i hjärnan." *Forskning och framsteg*, September 2011.

Svensson, Martina. "Min släktsaga." School paper, Klippans gymnasium, 2006.